ENIGMA RESEARCH UNLIMITED
INVESTIGATIONS OF THE PARA-NORMAL
P.O. BOX 462
REISTERSTOWN, MD. 21136

7

PARTICLES

PARTICLES

An Introduction to Particle Physics

MICHAEL CHESTER

DIAGRAMS BY ERICK INGRAHAM

MACMILLAN PUBLISHING CO., INC.
NEW YORK
COLLIER MACMILLAN PUBLISHERS
LONDON

Macmillan Publishing Co., Inc.
866 Third Avenue, New York, N.Y. 10022
Collier Macmillan Canada, Ltd.
Book design by Kathleen F. Westray
Printed in the United States of America

10 9 8 7 6 5 4 3 2 1

Library of Congress Cataloging in Publication Data
Chester, Michael.   Particles.
Includes index.
Summary: An introduction to the world of
atomic and sub-atomic particles, including positrons,
kaons, antiprotons, quarks, and others.
1. Particles (Nuclear physics)—Juvenile literature.
[1. Particles (Nuclear physics) 2. Nuclear physics]
I. Title.   QC793.27.C45   539.7′2   77–12352
ISBN 0–02–718240–1

CONTENTS

INTRODUCTION

1 THE universe we live in is a mysterious place. It stretches in all directions, apparently without end. And in this widespread world, we see a fantastic variety of things: earth, sky, and water; fire and lightning; flesh and fur; hoof, horn, and tooth; rock and sand; fish scale and bird feather; clouds and rainfall and rainbows; sun, moon, and stars.

Science has made it possible for us to see deeply into the nature of the world around us. But this ability has not dimin-

· · · 1

ished the mystery. Instead, scientists have found that the world is even stranger than it looks. Particularly when the smallest bits of matter are studied, a strange reality begins to take shape. The everyday world that we live in is made up of particles—extremely small bits of matter—that do very amazing things.

Any speck of matter can be called a particle. A grain of sand is a particle that is big enough to see with the naked eye, and obviously you could hold such a particle between your fingertips. The particles described in this book are far smaller than that. They are the smallest known bits of matter that make up the universe. They are the *fundamental particles*.

The laws of nature at this infinitesimal level are very different from what we know in our ordinary world. In the world of fundamental particles, one particle can be in several different places at the same time; particles can exist for fleeting instants of time in situations where there isn't enough energy available for them to exist at all; particles and "antiparticles" can annihilate each other, dissolving into pure energy.

These mind-boggling events are happening all the time—in our very bodies, in the air we breathe, and the rooms we walk through. In fact, nothing could exist if those strange events weren't happening, invisibly, among the fundamental particles. The real, "solid" world is shakier than you might think.

Consider this: A solid wall is really a thin cloud of particles, and you and I are such thin clouds of particles that we should be able to drift through a wall as easily as a cloud of smoke drifts through the air. If you were able to float through the wall into the next room, very few of the particles of your body would collide with particles of the wall.

What stops you from going through walls are the forces between particles. The wall doesn't resist you in a sluggish, lifeless way, as it seems to do. Instead, you are being stopped by an energetic force field, an invisible flow of energy through the empty spaces between particles. Your body is also a pulsating force field. You can't float through the wall because you can't push one force field through the other. However, you can push your way through the much weaker force fields of water or air.

People's inability to walk through walls is just one example of how our visible world is based on invisible, energetic events among particles that are almost too small to imagine. In many other ways, things that are familiar depend on this magical, almost impossible subworld.

How small are the particles that we are made of? That depends on which particles we look at. We can start with atoms, which are the basic building blocks of the world. Atoms are extremely small, by everyday standards. The period at the end of this sentence [a thin layer of ink about 1 millimeter (mm) in diameter] contains roughly three or four billion atoms. That is, there are about as many atoms in that dot as there are people living on the earth.

Since atoms are so small, a new unit of length is needed in order to talk about their sizes. The unit of length that scientists use is called an *angstrom unit*. It takes 100 million angstrom units to equal 1 centimeter (cm). An inch is about 254 million angstrom units. The angstrom unit is used throughout this book as a measure of length. It is abbreviated like this: Å.

Different kinds of atoms are of different sizes. But, in general, an atom measures about 1 or 2 angstrom units across. Even though they are extremely small, atoms are not the smallest things in existence. That is, they are not the fundamental particles. As we shall see, fundamental particles are much smaller than atoms.

To study atoms and the fundamental particles that atoms are made of, physicists deal with a reality that they can never directly see or feel. All of the scientific information that they obtain comes from measurements—simple measurements with rulers or clocks or scales, and subtler measurements that require some of the most intricate modern devices ever built. The deeper a physicist probes into the extremely small world of particles, the more indirectly the information comes to him. The wiggle of a needle on a dial of a sensitive instrument may be the only clue that a physicist has regarding a subatomic collision.

Since physicists depend on indirect clues relating to events far different from the events of everyday reality, they must form imaginary pictures of the things that are actually taking place. These pictures are usually called *models*. A good model in physics is one that is comparatively simple and compact, yet accounts for a wide variety of physical phenomena without contradicting itself.

A simple model of the nature of the universe is the idea that matter is made of atoms. A more advanced model is the idea that each atom is like a miniature solar system, in which smaller particles called "electrons" orbit around a central nucleus. This model, the "nuclear atom," was developed in 1911, and was already considered old-fashioned and incomplete by 1913 when a more advanced model was developed. Yet, for many purposes, this old model is still useful. It accounts for many different kinds of experimental results, and it does not break down until some very subtle details are considered. In fact, the common picture of the atom that most people are familiar with is this 1911 model.

The models that physicists use are like maps. Just as a road map provides clues on how to find a town or a street, the models of physics give us glimpses into the working of the physical world. We can use them to get some partial insight into the way our mysterious universe is put together.

ATOMS

2 TWO ideas were developed by the ancient Greeks that relate to particles. The first of these had to do with electricity. In about 600 B.C., Thales of Miletus discovered that a piece of amber (hardened tree sap), after it is rubbed with fur, attracts bits of hair and feathers and other very light objects. He suggested that a mysterious force was developed in the rubbed amber. That is the same mysterious force that we now call electricity.

Thales did not connect the idea of electricity with ideas about atomic particles. In fact, the atomic particle idea did not exist in his time. It wasn't until 140 years later that another Greek philosopher, Democritus, developed the idea of atoms.

The question that Democritus asked was this: If a small bit of matter (for instance, a stone, leaf, or piece of wood) is broken in half, and then broken in half again, to make smaller and smaller bits, do you finally come to the smallest possible bit of matter—something that can be split no further? Or could you theoretically keep going forever, breaking any bit of matter, however small it is, down into smaller bits?

Democritus felt that there was an end point, a smallest possible bit of matter. He realized that practical limits made it out of the question for him to cut objects down to such a fine level. But he believed that unbreakable particles existed and that all things were made of those particles. He gave these basic particles a name. He called them *atoms*.

Democritus considered that these atoms were uncaused, that is, they had always existed, and they could not be destroyed. He thought that atoms came in many different sizes and shapes, but were made out of the same basic stuff. The differences we perceive in substances would result from the way the atoms were connected together. One way of combining atoms would result in a metal—iron or bronze, for instance. Another way would result in water, another way in sand, and so forth. He also pictured that the atoms in a solid substance would vibrate or throb. But atoms of smoke or fire would fly about in free motion, colliding with one another to gradually form large objects, finally building up to fill the world and the objects in it.

Many of the ideas that Democritus had are close to modern atomic theory. Since he could not test his ideas about atoms through experiments, it is amazing that he could do as well as he did. He reasoned philosophically, perhaps operating partly from hunches and partly from logic. For instance, once he guessed that substances were made up of atoms, it was reasonable to assume that atoms of smoke or fire would move freely about (as the smoke and fire did) and that the atoms of a solid would be comparatively still. But, whatever combination of guesswork and logic he used, his results were impressive.

The ideas of Democritus had no lasting effects on other Greek philosophers in those early centuries. Aristotle, who summed up the thoughts of many earlier philosophers in his writings, dismissed the atomic idea as worthless. The original writings of Democritus were lost, and the only record that we have of his ideas is through Aristotle's writings and criticism. So the atomic idea had flashed up for a moment, deep in the past, and then was forgotten for many centuries.

The discovery of electricity by Thales was also preserved in Aristotle's writings. But neither Aristotle nor anyone else took any serious interest in pursuing the nature of this mysterious force. And nobody considered the possibility that electricity and the idea of atoms are connected in some way.

For more than 2,000 years, nobody did anything to continue the explorations that the Greeks had started into the nature of matter. The writings of Aristotle had been saved and copied and everyone took his opinions as the final word on scientific questions. Aristotle had declared that the atomic idea was wrong, and that closed the subject.

Finally, in the early 1800s, people began to be curious again about how the universe was put together. And, by this time, scientists had developed methods of experimentation. Instead of just thinking these things through as the ancient Greeks did, they could test their ideas out.

What experiments were possible, in the year 1800, to show whether matter was made of atomic particles or whether it could be broken down endlessly?

To illustrate the way that scientists approached this problem, let's consider an imaginary situation involving cups of coffee with sugar in them. Suppose you are presented with the following problem: You are given 20 cups of coffee to test. Different amounts of sugar were put in the cups. You have scientific equipment you can use to separate the sugar from the coffee, dry it out, and weigh it. The question is this: Was the sugar put into the coffee as *sugar cubes,* or was it *loose sugar* that was put in by the spoonful?

Sugar cubes or just loose sugar? How do you find out which was put into the coffee? It seems puzzling at first, because after sugar is dissolved in coffee, you can't tell by looking at it whether or not it came from sugar cubes.

But there is a way to solve this problem. Suppose that when you weigh the sugar that was in each cup you get the following results:

> 5 cups have exactly 1½ grams of sugar per cup
> 5 cups have exactly 3 grams of sugar per cup
> 5 cups have exactly 4½ grams of sugar per cup
> 5 cups have exactly 6 grams of sugar per cup

Now, notice a curious thing about this result. The five cups with the least sugar have 1½ grams each. The next five cups have exactly double that amount, 3 grams each. The third set of cups have 4½ grams of sugar each, which means that they have an amount that is exactly triple the amount of sugar found in the first cups. And the final batch of cups have exactly four times as much sugar as was found in the first cups.

This result would fit the sugar-cube idea very neatly. The first set of cups have one sugar cube each (assuming each cube weighs 1½ grams). The second set of cups have two cubes each. The third set have three cubes each. And the fourth set have four cubes each.

If loose sugar had been spooned into the cups, it could hardly be so exact unless someone measured the loose sugar very carefully. Usually, when loose sugar is used, the people who put it in the cups do so in a casual way, without exact measurement. They just scoop up a few spoonfuls or pour some sugar without measuring it.

If they put loose sugar into the cups in that way, then it is extremely unlikely that it could come out in the exact single, double, triple, and quadruple portions. Instead it might look something like this:

2 cups have ⅕ gram each
1 cup has ⅜ gram
1 cup has 1 gram
1 cup has 1⅕ grams
1 cup has 1⅝ grams
1 cup has 1⅞ grams
2 cups have 2 grams each
1 cup has 2¼ grams
1 cup has 2½ grams
3 cups have 2⁹⁄₁₀ grams each
1 cup has 3½ grams
1 cup has 4 grams
1 cup has 4⅓ grams
2 cups have 4⅞ grams each
1 cup has 6⅖ grams

These results could never come from sugar cubes. There are no simple proportions to fit the case of cups with one cube each, two cubes each, three cubes each, etc. (Let's assume that none of the cubes was split.) Random proportions indicate that loose sugar was put in the cups, just as exact proportions would indicate that sugar cubes were used.

SUBSTANCE: Any solid, liquid, or gas. All substances are made of atoms.

ELEMENT: Any substance that is made up of one kind of atom only.

COMPOUND: Any substance that is made up of two or more different kinds of atoms.

This imaginary experiment is very much like what the English chemist John Dalton did in the early 1800s to find out whether matter was made up of lumpy particles (atoms) or whether it was made of a loose substance that could be divided indefinitely.

Dalton carried out his research by studying chemical compounds and weighing the amounts of the various elements that were present. His results were very much like those for the sugar cubes. That is, elements were always found in simple proportions in different compounds. For instance, the portion of oxygen found in a sample of carbon dioxide (CO_2) gas is exactly double the portion of oxygen found in carbon monoxide (CO) gas. And chemical compounds in general have these simple proportions.

These results showed that matter was made of "lumpy stuff"—atoms—rather than "loose stuff" that could be endlessly divided. In this simple way, Dalton proved the atomic idea.

Dalton's proof of the atomic ideas of Democritus was not the only way that scientists were continuing the works of the ancient Greeks. In the 1600s, 1700s, and 1800s, the mysterious force that Thales had discovered was studied in detail, and an enormous body of knowledge began to develop concerning electricity.

The most basic facts about electricity are these:

1. Certain kinds of objects can become electrically "charged" when they are rubbed together. For instance, when glass is rubbed with silk, the glass and the silk both become electrically charged.

2. There are two kinds of electric charge. That is, when glass is rubbed with silk, the glass gets one kind of charge and the silk gets another kind. Eventually scientists gave names to these two kinds of charge. They could have called them anything: "Jack" and "Jill" or "Yin" and "Yang." What they actually did call them was *positive* and *negative*. It is important to remember that these words are just names and have no connection at all with the usual meanings of "positive" and "negative" in our language.

3. There are forces between electrically charged objects. Charges that are alike push each other away, and opposite charges attract each other. So, two positively charged glass rods would push each other away. Two negatively charged pieces of silk would also push each other away. But a positively charged glass rod and a negatively charged piece of silk would attract

each other, because they have opposite charges. For example, if you rub a glass rod with a small piece of silk and then put the silk on a table top and hold the glass rod near it, the silk is pulled toward the rod.

4. What is true of charged objects is also true of charged particles. Particles having like charges push each other away. Particles having opposite charges attract each other.

These are the most basic facts of electricity. Of course much more is known, involving the entire theory of electricity and magnetism as well as the fantastic modern world of electronics. But the simple fact of the two kinds of electrical charges and of the forces between electrical charges is the foundation of it all.

In the 1830s, experiments in "electrolysis" connected the atomic idea and the electrical idea into a single truth: Both positive and negative electric charges were to be found within each atom. In electrolysis, electric currents are sent through chemical solutions. Particles in the solution are sent into motion by these currents, in a way that shows them to be electrically charged.

To understand how electrolysis works, imagine an intelligent being from another universe looking down from outer space at a lake full of boats. His view isn't very clear, so he doesn't know whether the boats have sails. But he has a way of finding out; he has a huge wind machine. He turns on the wind machine and causes a breeze to blow across the lake. If the breeze makes the boats move, then they have sails. If the breeze doesn't make them move, then they don't have sails.

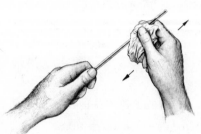

When glass is rubbed with silk, both the glass rod and the silk get opposite electric charges.

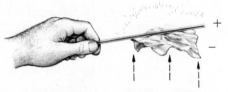

If the glass rod is held near the negatively charged piece of silk, the two will attract each other and the silk will cling to the rod.

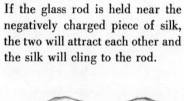

+ +

Opposite charges attract each other, but two positively charged glass rods will push each other away.

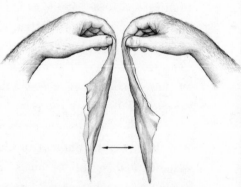

−   −

Two negatively charged pieces of silk will also push each other away.

Figure 1

· · · 14

In electrolysis, the "boats on the lake" are atoms in a chemical solution, the "sails" are electric charges, and the "wind" is an electric current. When an electric current is passed through a chemical solution, the atoms in the solution move if their sails are up; that is, if they are electrically charged.

Here is an example of electrolysis. Copper sulfate is a chemical made up of various elements, one of which is copper. But, looking at the blue crystals of copper sulfate, you see no trace of the bright red-brown metal that we call "copper." If some of these crystals are put into a small glass tank of water, they dissolve to form a transparent blue solution. Still no sign of copper.

Next, let's dip two rods of some stable material (such as carbon or platinum) into the solution and connect them to the positive and negative terminals of an electric battery. This makes a current flow through the solution. Soon a thin coating of red-brown metal begins to form around the negatively charged rod. Copper! The copper atoms have arrived at the negative pole just like the sailboats arriving at a pier. Since an electrical force made the copper atoms come to this pole, we know that they were charged while they were in the solution. Now, they give up that charge to become regular copper atoms, like sailboats that drop their sails when they dock.

One of the men investigating electrolysis in the 1830s was the English scientist Michael Faraday. Faraday had very little formal scientific education. Nevertheless, his experimental work and imaginative models are the basis for all that we know today about the subjects of electricity and magnetism.

Faraday gave the name *ions* to atoms that are electrically charged in chemical solutions. In the experiment described earlier, it is clear that the copper atoms are positively charged, since they are attracted to a negatively charged pole. In that same experiment there are also negatively charged ions that are attracted to the positive pole, where they react to release oxygen gas and acids. (The story of the negative ions is more complicated in this case, so they won't be discussed here.)

Faraday's experiments finally united the ancient ideas of Thales and Democritus. The atoms that matter is made of are held together by electrical forces.

But already atoms were starting to look a little complicated. If they had parts that were positively and negatively charged, then they weren't the smallest things in the universe. They had to be made up of even smaller things that carried those electrical charges. What were those parts of the atom and how did they behave?

RAISINS IN A BUN

3 IN the 1890s, physicists found that if they hit a piece of matter with a big enough jolt of electricity, the matter would give off invisible rays. These experiments usually involved sending electric arcs flashing between the two ends of a closed glass tube, as mad scientists like to do in science-fiction movies.

Other physicists discovered some substances, such as uranium and radium, that would give off invisible rays all by themselves, even if you didn't hit them with jolts of electricity.

Scientists gave these substances a name. They called them *radioactive* materials.

During the years that followed the discoveries of the new rays, careful studies were made. One kind of wavelike ray was so energetic that it could go through thick objects. These energetic rays were named *X rays*. Other rays proved to be streams of particles. The physicists who were investigating these particle rays called them *beta rays* (β rays).

Beta rays bend when passed near a magnet. From this fact, it was clear to the physicists that the beta particles were electrically charged, and the direction of the bend indicated that they were negatively charged. Additional experiments showed that the beta particles were fantastically small. It would take nearly 2,000 of them to equal the mass of the smallest known atom.

Also, the beta particles that came from different substances seemed to be identical. It didn't matter, for instance, whether the beta rays were given off by a lump of radioactive substance or by a strip of platinum metal struck by an electric arc. The beta particles were the same.

The fact that beta particles were so small and that the same beta particles came from different kinds of substances was very interesting. It seemed that these particles were from a subatomic level, that they were, in some way, pieces of atoms. Perhaps these were the fundamental building blocks of the universe. After a while, physicists stopped calling them beta particles. Instead, they called them *electrons*.

The discovery of the electron was made in 1897 by the English physicist J. J. Thomson. You might think, from his having made such an important discovery, that Thomson was very good at working with his hands in the experimental laboratory. Actually, just the opposite was true; he was awkward and uncomfortable when it came to working with his hands. So, when Thomson carried out experiments, he had assistants who handled the equipment. But Thomson's imagination and logic shaped the experiments and the discovery that resulted.

Because of the way electrons streamed out of all substances, Thomson was convinced that every kind of atom had to be made of electrons. Yet, it was obvious that electrons could not be the only parts of the atom. First, there was the electrical situation. As we have seen, electrons carry a negative charge. But atoms are electrically neutral. We know this because matter, made out of atoms, is usually neutral, rather than electrically charged. So, every atom had to contain electrically positive material too. The negative electrons and the positive material would balance each other to leave the atom electrically neutral.

Also, the electrons were far too lightweight (low *mass* in scientific language) to be the main parts of the atom. The positive material would have to contain most of the mass.

From these facts, Thomson formed his picture of what atoms must be like. He saw an atom as a ball of positively charged matter with the much smaller negative electrons scattered through it, like raisins in a bun.

This picture fitted many experimental facts. For instance, even though matter usually is electrically neutral, it can be charged in various ways, such as through friction, or through chemical reactions. Using Thomson's model, this charging process can be seen as a stripping of electrons from atoms. The elec-

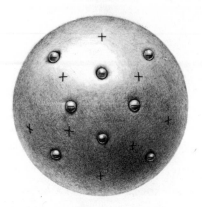

Figure 2

Thomson's model of the atom

Thomson saw an atom as a ball of positively charged matter with negatively charged electrons embedded in it, like raisins in a bun.

trons that are torn away may appear in different forms: as beta rays coming from radioactive material, arcs of electricity jumping between platinum electrodes, electrical current flowing through a copper wire, and countless other forms. The atoms that lose electrons are no longer electrically neutral. They are left with a positive charge.

In some cases extra electrons can be packed into atoms, leaving the atoms negatively charged. But, whether objects are positively or negatively charged, the charging process can be seen as evidence of the motions of electrons out of some atoms and into others. That part of Thomson's model still holds true. However, the way in which the electrons fit into the atom is far different from the way Thomson imagined it.

The raisins-in-a-bun idea was not destined to last very long. Thomson's model was only a rough picture of what an atom is like. In some situations, his idea did explain the behavior of matter. In many other situations, his idea turned out to be useless. His model rapidly became obsolete.

THE
NUCLEAR
ATOM

4 ANOTHER kind of radiation discovered around the turn of the century was *alpha* (a) radiation. Like the beta rays, the alpha rays were found to be made up of particles. However, the alpha particles were much larger than electrons. They were obviously large pieces of atoms, and they were positively charged. At the time it seemed that they might be the positive parts of the Thomson atom.

In 1911, scientists decided that it would be interesting to

use the big alpha particles to bombard other atoms. They would send these atomic cannonballs crashing into targets, to see what would happen. Of course all this was on a very small scale. An invisible ray would be aimed at a thin piece of gold foil— much thinner than the aluminum foil that is wrapped around a piece of gum. As they passed through the foil, the alpha particles in the ray would hit gold atoms.

This bombardment was a way of "looking into" the insides of atoms. An atom is so small that something of atomic size or smaller has to be used in order to find out what it's made of. If your eyes were closed, you could investigate the inside of a walnut shell with one finger, but you couldn't do it with your fist or with your head. Atomic bombardment is based on the same idea. Only something extremely small can be used to probe into the atom.

The leader in this new research was Ernest Rutherford, a big, loud-voiced, mustached Australian. His research, carried out in Canada and (later) at the Cavendish Laboratory in England, was of tremendous importance in the study of atomic structure.

Working under Rutherford's bossy, determined leadership, Ernest Marsden set out to bombard thin layers of gold foil with radiation in the form of alpha particles.

What Rutherford had in mind was this: When the positively charged alpha particles came near charged parts of the gold atoms, they should swerve slightly from their paths. Looking at the pattern of how the alphas swerved was a chance to see what the insides of the Thomson "raisin bun" looked like. Instruments would detect the swerving alphas after they went through the foil.

Rutherford and Marsden weren't sure what to expect. They were like anyone would be upon poking into the inside of something that nobody had ever poked into before.

What they found was this: Almost all of the alpha particles went smoothly through the foil, swerving very slightly or not at all. Only an occasional alpha veered sharply from its original path, sometimes even bouncing straight back from the foil.

This was very different from what they expected. Suppose you throw tennis balls at a wall, and most of them go straight through the wall, but once in a while a ball bounces back to you.

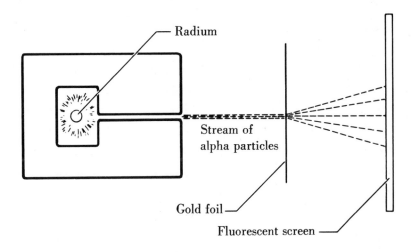

Radium

Stream of alpha particles

Gold foil

Fluorescent screen

Figure 3a

Rutherford's experiment

Schematic diagram of the experiment in which Rutherford discovered the nuclear atom

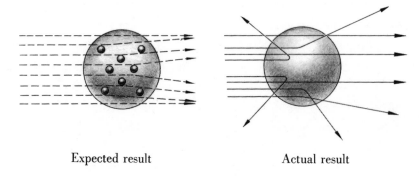

Expected result   Actual result

Figure 3b

You know that there is something strange about that wall; it acts like a normal wall only part of the time. That was exactly what Rutherford and Marsden found when they sent the alpha particles into the gold foil.

By studying the pattern of the scattered alphas, Rutherford found that they must be getting scattered by concentrated bits of positively charged matter. Most of the space around these positive centers had to be empty or thinly "sprinkled" with negatively charged bits of matter. The alphas that went straight through the foil were crossing these empty parts of atoms. The few alphas that veered were those that came close to centers of positive charge.

This "empty" atom with all its positive charge packed into the center reminded Rutherford of something familiar. But what? Then he realized. Of course! The solar system! The sun is in the center, and the planets circle through the empty space around the sun. Atoms must be like that too. The center of positive charge is the *nucleus* of the atom. And the negatively charged electrons orbit around the nucleus. That was what the scattering experiment seemed to say.

And that is how Rutherford's nuclear model of the atom was first thought of. It was a useful model in 1911. And it is still a useful model. Atoms are still pictured as looking like miniature solar systems, with electrons orbiting around a nucleus. Furthermore, many physical phenomena can still be explained through the use of this model.

But, even in 1911, there were serious difficulties in the

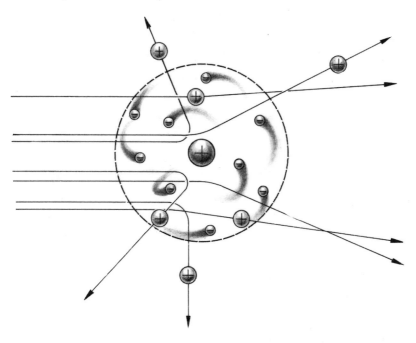

Figure 3c

How alpha particles are deflected by a center of positive charge in a nuclear atom

Rutherford idea. The theory of electricity and magnetism was very definite about what negatively charged particles would do if they were orbiting around a positively charged nucleus. Since particles having opposite charges attract each other, the electrons would gradually lose energy and spiral inward. And, as they spiraled inward, they would give off a rainbow of light.

But atoms do not give off rainbows of light. What atoms actually do is something very different.

All of this has to do with something called a "spectrum." When sunlight shines through droplets of water on a rainy day, you see a rainbow spectrum. The same thing can be done in a scientific laboratory when sunlight shines through a wedge-shaped piece of glass called a prism.

If a solid or liquid or a dense gas is heated until it glows, its light does what sunlight does. Shining through a prism, it spreads into a complete rainbow spectrum. In other words, the spectrum is a wide band of colors, having all the colors of the rainbow. This kind of spectrum starts with the short wavelengths of violet light (with a wavelength of about 4000 Å), followed by blue, green, yellow, and orange, as wavelengths get longer, and ending with red. Red light has the longest wavelength of visible light—just under 8000 Å, or about double the wavelength of violet light. (The wavelength of a light wave is the distance between peaks of the rippling wave.)

A very thin gas, when it is heated to the glowing point, produces a *line spectrum*. This is a spectrum in which lines of colored light appear at certain definite wavelengths. For example, if hydrogen gas is heated to a high temperature so that the hydrogen molecules break down into single atoms of hydrogen, the spectrum of atomic hydrogen can be obtained. This spectrum, under low pressure, consists mainly of four very bright lines—red, blue-green, blue, and violet—and some fainter violet lines. On the other hand, glowing sodium gas under low pressure has a spectrum of two yellow lines.

Each kind of atomic element has its own characteristic line spectrum. The spectrum of each element is totally dark except where the lines appear. In other words, each kind of atom radiates light only at certain definite wavelengths and gives off no light at all at other wavelengths. The line spectrum of each element is the unique signature of that element.

Atoms in a high-pressure gas or a liquid or solid are packed more tightly together, and they disturb each other's motion and energy states. Therefore these atoms are no longer so "sharply tuned" to certain exact wavelengths. Instead, they radiate over all wavelengths, resulting in a continuous "rainbow" spectrum instead of a line spectrum. But that is only a distortion of the way individual atoms radiate energy. The line spectrum shows the true picture.

And this brings us back to the trouble with the Rutherford model. Atoms give off line spectra; according to the Rutherford model, they would only produce rainbows. So something was wrong with the model.

Actually, the wrong kind of spectrum was a minor problem when another result was taken into consideration. According to the laws of electricity and magnetism, the negatively charged electrons would plunge into the positively charged nucleus, and the atom would collapse. If the Rutherford model were true, all matter would be unstable and should collapse immediately.

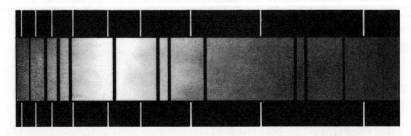

Sunlight

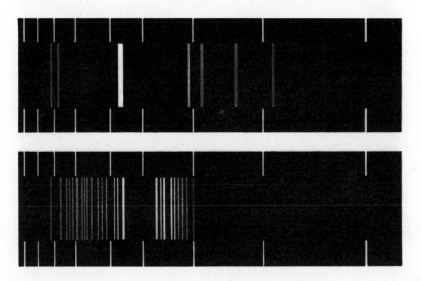

Above: helium spectrum   Below: neon spectrum

Figure 4

The continuous spectrum of the light from the sun is crossed by fine dark lines. These represent a rainbow spectrum. In the illustration only the main lines are shown.

When the light from a low-pressure luminous gas goes through a prism, it is found to produce a line spectrum.

THE IMPOSSIBLE ATOM

5 THE problems with Rutherford's model of the atom were resolved in a surprising way by a young Danish physicist named Niels Bohr. In 1912, Bohr stated some rules for the way electrons orbit around the atomic nucleus. What made his approach especially interesting was that he did not try to justify his rules or find reasons for them. The rules made very little sense, when they were put up against firmly established theories of physics. In effect, Bohr was saying, "Here are some rules that seem impossible, but they do describe the way atomic systems appear to operate, so let's use them."

Bohr started by assuming that orbiting electrons do *not* spiral into the nucleus. This contradicted everything that was known about electricity and magnetism, but it fit the way things obviously were.

Then Bohr stated his two rules for what does happen.

RULE 1. Electrons can orbit only at certain allowed distances from the nucleus.

Consider the hydrogen atom, for example, which has only a single electron orbiting around the nucleus. Bohr's calculations showed which orbits were possible. Figure 5 shows the first five of these allowed orbits. The first orbit lies just over half an angstrom unit from the nucleus (0.529 Å). The second allowed orbit is a little more than 2 angstroms from the nucleus (2.116 Å).

Although Fig. 5 shows only the first five orbits, there is no end to the number of orbits that are theoretically possible. For example, the hundredth Bohr orbit for the hydrogen atom would lie 10,000 times farther away from the nucleus than the first orbit, at a distance of 5290 Å.

However, extremely distant orbits, such as the tenth, twentieth, or hundredth orbit, are unlikely. The chances are that an electron in a very remote orbit would be lost from the atom. That is, another atom would capture it, or a passing ripple of electromagnetic energy would set it adrift as a "free electron" moving through the space between atoms. So the more important orbits,

Orbit	*Distance from Nucleus*
1	0.529 Å
2	2.116 Å
3	4.761 Å
4	8.464 Å
5	13.225 Å

Figure 5

Bohr orbits for hydrogen atom

those that play the main part in producing the line spectrum of an atom, are the innermost orbits.

It is rather a strange rule that electrons can occupy only certain fixed orbits. It means that most orbits would be impossible. A hydrogen electron could not orbit at 0.250, 1.000, or 2.150 Å; the only allowed orbits are those listed in Fig. 5.

This is quite different from the way that objects act in our everyday world. Suppose that a ball tossed across a room could follow only two or three certain paths, instead of the hundreds of different paths it actually can follow. It would be as if the room had invisible tracks guiding the ball. So, Bohr's rule says that electrons act as if the space around the atomic nucleus has invisible tracks. But Bohr gave no reason for this strange situation.

This brings us to the second rule that Bohr stated.

RULE 2. An atom radiates energy when an electron jumps from a higher-energy orbit to a lower-energy orbit. Also, an atom absorbs energy when an electron is boosted from a lower-energy orbit into a higher-energy orbit.

In other words, electrons leap from one allowed orbit to another as atoms radiate or absorb energy.

The outer orbits of the atom are higher-energy orbits than the inner orbits. So, if an electron drops from orbit 2 to orbit 1, light is given off. On the other hand, if light of the right energy strikes the atom, it can boost an electron from orbit 1 up to orbit 2. In that process, the light is absorbed.

Interestingly, the wavelengths of light that are found in the spectrum of hydrogen match the energy gaps between the different orbits. (Wavelength is related to energy. Shorter wavelengths of light mean more rapid vibrations and greater energy.) For instance, the blue-green line in the hydrogen line spectrum is caused by electrons dropping from the fourth orbit to the second orbit. Figure 6 shows how each line in the spectrum results from a particular jump.

Notice that all the jumps in Fig. 6 are jumps from higher orbits to orbit 2. The lowest-energy jump is the one from the third orbit to the second, and this jump causes the red line at 6563 Å. Finally, there is a cluster of lines at the violet end of the spectrum, caused by electrons jumping from distant, outer orbits down to the second orbit.

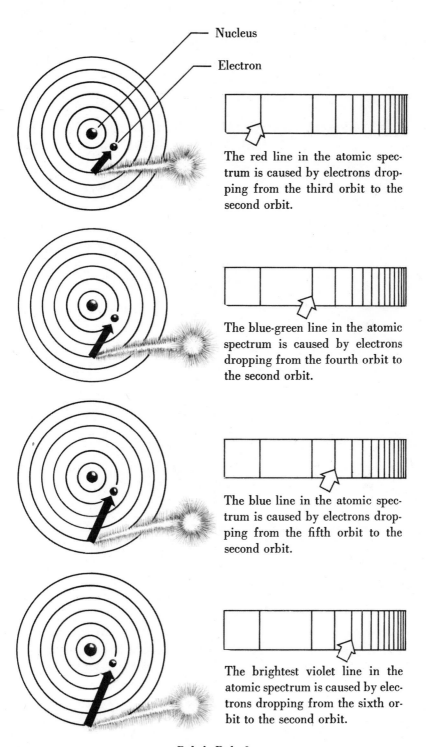

Nucleus

Electron

The red line in the atomic spectrum is caused by electrons dropping from the third orbit to the second orbit.

The blue-green line in the atomic spectrum is caused by electrons dropping from the fourth orbit to the second orbit.

The blue line in the atomic spectrum is caused by electrons dropping from the fifth orbit to the second orbit.

The brightest violet line in the atomic spectrum is caused by electrons dropping from the sixth orbit to the second orbit.

Bohr's Rule 2

Figure 6

In the case of hydrogen atoms, only jumps down to the second orbit cause spectral lines in the visible part of the spectrum. Jumps to the first orbit cause ultraviolet radiation (waves that are shorter than light waves), while jumps down to the third, fourth, and fifth orbits cause infrared radiation (waves that are longer than light waves).

Bohr's definite orbits and the way the electrons jumped between them got rid of the old picture of electrons spiraling into the nucleus. It also got rid of the requirement that atomic radiation should be a continuous rainbow of light, and accounted for the line spectra.

It was all very strange. Bohr's bold, imaginative ideas had led to something that worked very well. But neither Bohr nor anyone else could understand exactly how it worked.

QUANTUM NUMBERS

6 NIELS BOHR first proposed his "impossible atom" in 1912. But by the mid-1920s, Bohr's model of the atom was gradually running into trouble. Calculations based on the model gave fairly good results for various kinds of atoms. Hydrogen, helium, sodium, and all the other elements had bright line spectra very much like what Bohr's model predicted they would have. Very much, but not exactly. . . .

To understand how the actual spectra differed from the

predictions, we should look at some of the fine details that Bohr and other scientists added to the model as time went along.

At its simplest, the Bohr model depended only on which orbits the electrons were traveling in. An electron jumping from the fourth allowed orbit to the second allowed orbit of a hydrogen atom would produce light at a certain definite wavelength. And that was the whole story.

The simplicity of this model was somewhat deceptive. To begin with, a hydrogen atom is the simplest of all atoms, with only a single electron in orbit. Would the Bohr model work so well for more complex atoms? Would it work for the helium atom with two orbiting electrons? Or, what about the fluorine atom with nine electrons, or the calcium atom with twenty electrons in orbit?

The Bohr model worked only roughly in these cases. It did accurately predict the wavelength of most of the spectral lines. But many of the predicted lines were double lines or multiple lines. They were made up of light of slightly differing wavelengths grouped in little clusters very close to the predicted wavelength. In other words, the predicted spectral lines were splitting into several lines.

Spectral lines would also split into multiple lines (even in the case of hydrogen) when a strong magnetic field was operating in the glowing gas. There was no basis for this magnetic splitting in the original, simple version of the Bohr theory.

In addition to these splittings of spectral lines, there were other inconsistencies. In some cases, lines that were predicted by the Bohr theory did not appear, and lines that were not predicted did appear.

Bohr and a German theoretical physicist named Arnold Sommerfeld refined the original Bohr model in order to try to explain these variations. And they did fairly well with their improved model. According to the Bohr–Sommerfeld model, not only do electrons travel in special orbits, but the orbits have different shapes and there are different ways each orbit could tilt in the presence of a magnetic field.

Some orbits are perfectly circular, others are elongated into an ellipse. In some cases, an electron could even swing back and forth through the nucleus in a straight line. This was astonishing. It was like a bird flying back and forth through a mountain. Nobody could make any sense out of it.

Whatever the shape of the orbit, it could be tilted at different angles, just as someone can tilt a hat at various angles on his or her head, no matter what the shape of the hat brim. In the case of orbiting electrons, the orbits are tilted at various angles to the direction of a magnetic force field.

There are only certain allowed shapes and only certain allowed tilts that an electron orbit can have. For example, the fourth orbit in a hydrogen atom can have only three possible shapes and seven possible tilts.

When electrons are allowed to change the shapes or tilts of their orbits, in addition to jumping from outer orbits to inner orbits, there are more possibilities for different spectral lines to appear. These possibilities brought the model into closer agreement with the experimental data.

In general, the orbit of an electron can be described by *quantum numbers*. The three quantum numbers that we have been discussing so far are: orbit number (n), orbit shape (l),

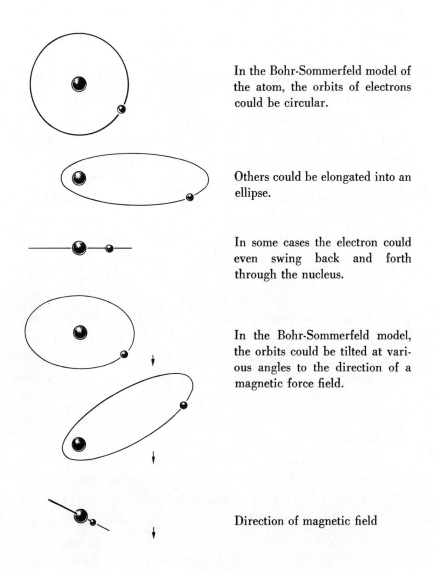

In the Bohr-Sommerfeld model of the atom, the orbits of electrons could be circular.

Others could be elongated into an ellipse.

In some cases the electron could even swing back and forth through the nucleus.

In the Bohr-Sommerfeld model, the orbits could be tilted at various angles to the direction of a magnetic force field.

Direction of magnetic field

Figure 7

The "improved" model of Niels Bohr and Arnold Sommerfeld

orbit tilt (m). Some years later, a fourth quantum number was added, when the young Austrian physicist Wolfgang Pauli predicted that an electron should spin like a top while it is orbiting around the nucleus. The electron can spin in either of two directions. So, electron spin can be described by a fourth quantum number: electron spin (s).

In 1924, Pauli announced a rule governing the behavior of electrons within the atom. If an electron has a certain set of quantum numbers (for instance, $n = 5$, $l = 2$, $m = -1$, $s = \frac{1}{2}$), then no other electron in that atom can have the same set of quantum numbers.

The numbers in the example above describe an electron in the fifth orbit, with orbit shape 2, a tilt of -1, and a spin of $\frac{1}{2}$. It's not obvious what a shape of 2, or a tilt of -1, or a spin of $\frac{1}{2}$ is, except to an advanced student of physics who has learned the full details of this theory. While it isn't practical to go into that kind of detail here, it is important to realize that this set of numbers (5, 2, -1, $\frac{1}{2}$) completely describes the motions of an electron. Any other electron in that same atom is required, by this fundamental law, to have a somewhat different motion. Another electron could be the same in three of those motions, as long as it was different in the fourth one. For instance, its quantum numbers could be ($n = 5$, $l = 2$, $m = -1$, $s = -\frac{1}{2}$), which would be an electron moving in an identical way but with a reverse spin, shown by the minus sign in front of the $\frac{1}{2}$. On the other hand, another electron in that atom might have an entirely different set of quantum numbers such as ($n = 3$, $l = 0$, $m = 0$, $s = \frac{1}{2}$).

The rule that requires each electron in an atom to move dif-

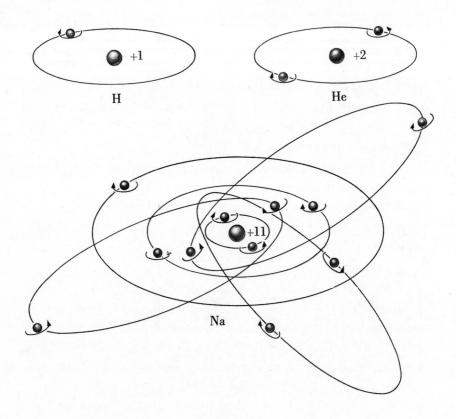

Figure 8

According to the Pauli exclusion principle, only two electrons of opposite spin may occupy the same orbit. The illustration shows how the atoms of hydrogen (H), helium (He)
and sodium (Na) are constructed on this basis.

ferently is called Pauli's exclusion principle. It is an important principle to this day, and it has even outlived the Bohr–Sommerfeld model that it was designed for.

The various motions of electrons, bounding from an orbit of one kind to an orbit of another kind, made the Bohr–Sommerfeld model a very rich and intricate sort of thing, like some kind of ingenious toy. But, for all its beauty and cleverness, it was largely a patch-up job. In fact, it went too far in all the possibilities that it provided. It predicted more lines than existed in the actual spectra of the various atomic elements.

Also, the Bohr–Sommerfeld model didn't connect in any way to the different intensities of brightnesses of the observed spectral lines. A bright line was one that resulted from many electrons making the same orbital jump. A dimmer line would result from fewer electrons in the sample making that particular jump. So, the varying brightnesses of spectral lines suggested that some electron jumps were more likely to happen than others. But nothing in the Bohr–Sommerfeld model had anything to do with how likely the chances of these jumps were.

For these reasons, the Bohr–Sommerfeld model was starting to look wobbly. And, with strange ideas about the nature of light pouring in from all sides, the trouble was getting worse all the time.

LIGHT

7 THE speed of light is fantastically great. By modern techniques of measurement, it has been estimated at 186,000 miles per second, or 30 billion centimeters per second.

Picture a distance of 30 centimeters (about one foot); a light ray covers that distance in one billionth of a second. Think of it another way: Put two mirrors half a foot apart; in a thousandth of a second a light flash could go back and forth between the two mirrors a million times. It is this enormous speed that makes light seem as if it is instantaneous. Only scientific experiments prove that it takes time for it to travel.

More than a hundred years before John Dalton proved that matter is made up of particles called atoms, another great scientist, Isaac Newton, was trying to prove that light was made of particles. Newton pictured a lamp or a fire or a star to be sending out constant streams of tiny particles. The reflection of light from a mirror or any other object would be described as particles of light bouncing off the surface. But another leading scientist of that time, Christian Huygens, disagreed with Newton. Huygens felt that light was a kind of wave motion—a ripple of energy traveling through air or water or empty space.

Since we have been discussing light as a form of wave motion, let's talk about waves in general. Exactly what is a wave?

By holding a rubber hose taut, with one end attached to a faucet and the other end in your hand, and giving a sharp downward jerk to your end of the hose, you can send a rippling wave along the hose. This clearly visible wave travels from your hand toward the faucet. What travels is only a disturbance, a ripple of energy. The same thing is true of the ripples that are caused by a pebble tossed into a pond, or the vibrations that travel along a steel rod when one end of the rod is tapped.

In the case of the steel rod, it is fairly obvious that blobs of steel aren't traveling from one end of the rod to the other. The same thing is true in the less obvious case of ripples in a pond or waves in the ocean. Water isn't traveling across the surface of the pond or sea; instead, an up-and-down disturbance travels along the surface, making the water move up and down as the disturbance goes past. The cause of the disturbance

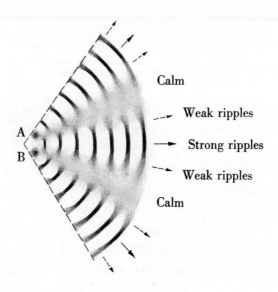

Calm

- - ▸ Weak ripples

⟶ Strong ripples

- - ▸ Weak ripples

Calm

Figure 9

The motion of a two-pronged fork causes two wave fronts to move out from positions A and B. The wave fronts interfere, causing strong ripples in some directions, weak ripples in other directions and calm water in still other directions.

You can see this effect by vibrating a two-pronged fork in a well-lit basin of water. Watch the shadows on the bottom of the basin to see where the ripples are strong or weak.

is some outside force, that is, energy is added when the pebble is dropped in the pond, or the wind sweeps across the ocean, or the steel rod is tapped, or the hose is shaken.

One thing that all kinds of wave motion have in common is something called an *interference pattern*. An interference pattern can be created by any kind of wave motion, for instance, by ripples in a pond, or waves in the sea.

In Fig. 9, wave fronts are being formed in a basin through the rapid motion of a two-pronged fork. Where the two wave fronts cross each other, there is interference. That is, each set of waves interferes with the motions of the other set. Where the waves are "in phase" with each other, so that their crests are in the same places, their energies add together, so that a bigger, more energetic wave motion results.

When waves arrive "out of phase" with each other, so that the peaks of one wave and the troughs of the other wave come together, the waves cancel each other, leaving a compara- tively calm place in the water (see Fig. 10).

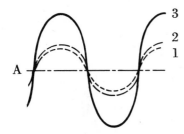

Figure 10a

Waves "in phase"—where waves 1 and 2 combine to make wave 3

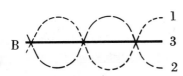

Figure 10b

Waves "out of phase"—where waves 1 and 2 cancel to make calm area 3

Light also forms an interference pattern. This is shown very clearly in the "double-slit" experiment. Light is sent through a pair of slits (shown in Fig. 11). A wave front spreads out from each slit. The two wave fronts interfere with each other, just as the wave fronts do in the water, and an interference pattern of bright and dark bands is formed on a screen. The fact that light can make interference patterns is strong proof that light is made up of waves.

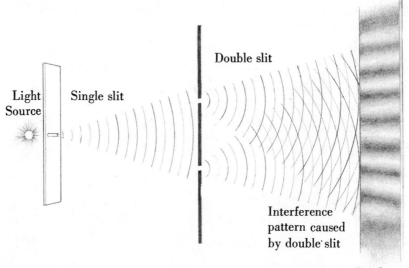

Figure 11

Young's experiment

But waves of what? Ripples in a pond or waves in the sea are waves of water. The waves that travel down a hose are waves of rubber. Sound waves are waves of air or metal or water, whatever substance the sound is traveling through. On the other hand, light waves travel through empty space even more easily than through air or water or glass. What is the "stuff" that ripples to make light waves?

The story of how this question was answered is a long one, involving many of the greatest and most imaginative physicists who ever lived. It is a story that would take us too far off the track of particles. So let's skip the history and go directly to the answer. What these physicists found was that light waves are ripples in electric and magnetic force fields. This led to the general idea of *electromagnetic waves*.

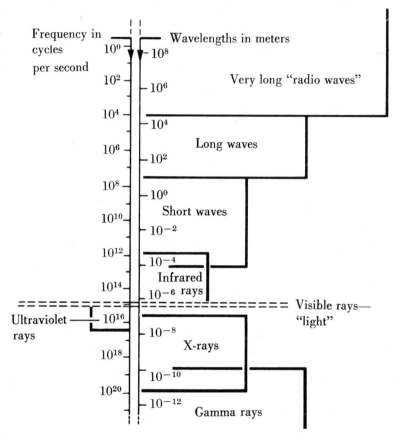

Figure 12

The electromagnetic spectrum

If light waves are electromagnetic waves of certain lengths (between 4000 and 8000 Å), then there should be electromagnetic waves that are shorter than light waves and also electromagnetic waves that are longer than light waves. So, the research into light didn't lead just to light waves, it led to a whole family of waves.

Shorter than light waves	Longer than light waves
gamma rays (γ rays)	infrared rays
X rays	radar waves
ultraviolet rays	radio waves

However, by the early part of the twentieth century these electromagnetic waves were heading for trouble. They were breaking on the rocky coast of particle physics, and strange things were happening to them there. In fact, it wasn't obvious whether these waves were really waves at all. Perhaps they were particles.

The idea that light rays could be streams of particles was astonishing. It was true that there once had been doubt as to whether light was made of waves or particles. But interference patterns and other definite proofs over the past hundred years had long ago decided the question. How could the particle idea be coming back into the area of light rays?

However, it was coming back. Many experiments showed that light waves are made of particles. In 1901, Max Planck, professor of theoretical physics at the University of Berlin, showed that the atoms of glowing objects give off light in separate bursts of energy. These energy bursts were named *quanta*. A single energy burst was called a *quantum*. These separate bursts seemed more like particles than waves.

Other experiments around the turn of the century had to do with collisions between light rays and electrons. In these collisions, the electrons did not act at all as if they were being hit by waves, they acted as if they were being hit by particles. Finally, these particles of light were given a name. They were called *photons*.

However, the issue wasn't closed. The various proofs that light was a form of wave motion were still strong. It seemed as if there was something both wavelike and particlelike about light rays.

Physicists were totally perplexed. It seemed impossible for something to be both a wave and a particle. The situation was much more puzzling than it had been over two hundred years earlier. Then there had been no good proof for either the particle or wave theory of light. Now there were good proofs of *both* theories.

MATTER WAVES

8 IN 1924, a year before Pauli discovered the exclusion principle, a Frenchman named Louis de Broglie was thinking about particles of matter. He was wondering why everyone was so sure that they really were particles. The nature of light was acknowledged to be a mystery; sometimes light seemed to be made of waves and sometimes it seemed to be made of particles. Why, then, should matter be any different? Could particles of matter, such as the electron, actually be seen as waves of matter?

De Broglie wasn't sure. But the question interested him.

There was one clue suggesting that waves of matter should exist. The clue came from Albert Einstein's special theory of relativity (1904). Einstein's famous equation $E = MC^2$ showed that mass is a form of energy and that matter can be converted into free energy; for instance, into a huge burst of electromagnetic waves. This possibility seemed to suggest a certain "waviness" to the nature of matter.

In a few neat, simple equations, de Broglie described what matter waves would be like if they existed at all. One of the things that he did was to calculate the wavelength of an electron. For an electron moving at 1% the speed of light, this wavelength would be about two angstrom units—an extremely short wavelength. In fact, it was such a short wavelength that it wouldn't be easy to detect the "electron waves." Perhaps that was why the wave nature of the electron had never been noticed—if the electron had a wave nature.

Not long afterward, the question was unexpectedly answered by two American scientists who weren't even thinking about matter waves. Their names were Clinton Davisson and Lester Germer. They were carrying out research at Bell Telephone Laboratory in New York, bombarding a piece of metal with a stream of electrons. The original purpose of their experiment is of no particular concern here, for an accident happened that changed everything.

The accident was the explosion of a flask of liquid nitrogen, which scattered bits of their equipment all over the lab. Fortunately, nobody was injured, and the two physicists set to work picking up the pieces and setting up their experiment again. Davisson and Germer didn't realize it, but the explosion had changed the microscopic structure of the piece of metal they were working with. Its new structure happened to be an orderly arrangement of metallic crystals that was ideally structured for the detection of very short waves.

Not realizing the change, and not realizing, in any case, that electrons could be anything but particles, they set up their high-voltage equipment and continued to bombard the lump of metal with a beam of electrons. Then they stopped to examine the metal under a microscope, an examination that was a key part of their original experiment.

What they saw under the microscope was something that they had never expected to see. A pattern of light and dark bands stood out with startling clarity. It was an interference pattern. The electrons had apparently hit the surface of the metal as waves, not as particles.

Davisson and Germer followed up on this experiment. They found out about de Broglie's theory, and they continued to experiment with the wave nature of electrons. Their results were conclusive. Electrons were waves.

However, there was a wealth of other data, including photographs of electrons in cloud chambers (where high-speed particles leave visible vapor trails) to prove that electrons were particles. Matter was now in the same condition that light was in. There was excellent evidence to prove that matter was made of particles—or of waves—depending on what kind of experiment was carried out.

In 1926, a twenty-three-year-old Austrian physicist named Erwin Schrödinger had an interesting idea: Why not go all the way with this new picture of electrons as both waves and particles and try to form a model of the atom on that basis?

Schrödinger treated an orbiting electron as if it were accompanied by a matter wave. Then he tried to "fit" this wave to the orbit. In most cases, as in Fig. 13a, the wave would not fit. That is, it would get out of phase with itself, overlapping and canceling itself out. But some orbits were exactly right, and the matter wave would fit very neatly as in Fig. 13b.

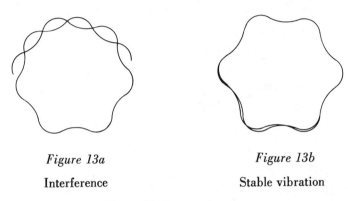

Figure 13a

Interference

Figure 13b

Stable vibration

Two orbital-wave situations

The interesting thing about this result was that the orbits where the waves fitted were the exact orbits of the Bohr model. Therefore, Bohr's allowed orbits no longer had to be the result of arbitrary rules. Instead, they were a natural result of matter waves. They were the orbits where the matter waves would fit without getting out of phase.

Schrödinger's idea was powerful because it combined the ideas of the electron as particle and wave. Starting with the speed of the particle, he could use a formula to find the wavelength of the matter waves. From the wavelengths of the matter waves, he could find the allowed orbits. In changing orbits, the wave-electron would have to give off light representing the difference in wavelength between the two orbits. These were the wavelengths that actually appeared in the line spectra. Everything fitted together beautifully.

With one exception. Schrödinger's model described the behavior of particles orbiting around a nucleus. But what about free particles, electrons, or other particles moving through interatomic space? How could a freely moving particle be combined with matter waves?

Schrödinger imagined that a large number of matter waves, of many slightly different wavelengths, were moving along. In general, a collection of waves of this sort, overlapping each other as in Fig. 14a, would cancel each other out. But, occasionally the waves would overlap as in Fig. 14b, with all the waves having their peaks together at some one point (P), forming a combined larger peak at that point. Everywhere else they would cancel out. The result would be the little "wave packet" of Fig. 14b. In fact, this packet, built out of the separate matter waves, would act like a particle. Perhaps this was what a particle was—a *wave packet*. The waves could show wave characteristics such as interference patterns, and the packet could collide with other packets as a particle. Schrödinger's idea could solve the double identity of the photon, the electron, and every other thing that had both wave and particle characteristics.

However, this beautiful solution had a serious flaw. The matter waves with their slightly different wavelengths would also have to have slightly different speeds, and they would drift out of phase with each other. In a very small fraction of a second, they would overlap each other randomly, as in Fig. 14a. Then there would be no more wave packet. So, particles could not be stable and could not continue to exist.

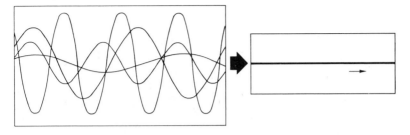

Figure 14a

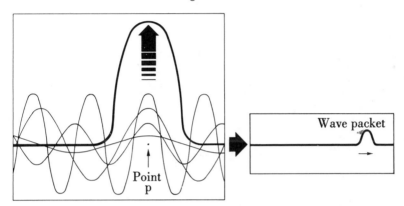

Figure 14b

Schrödinger's idea that combined the electron as particle and wave in a "wave packet"

The fact that Schrödinger's *wave mechanics* seemed to break down for free particles threw doubt on his entire approach. It was really too bad, because his approach seemed very close to being an answer for all sorts of problems. It made sense out of the Bohr atom with its quantum conditions that had previously seemed like arbitrary rules. It also offered a compromise between the wave and particle ideas. These were great accomplishments. The fact that wave packets fell apart seemed to ruin everything. Yet, wave mechanics was very promising. Perhaps there would be a way to save it after all.

WAVES
AND BLOBS
OF CHANCE

9 THERE was one question Schrödinger never answered in his theory of wave mechanics. The question was "waves of what?" What was the stuff that was rippling to make matter waves? Even though Schrödinger didn't answer that question, he gave this stuff a name, or at least a symbol. He used the Greek letter ψ (psi) to stand for the stuff that the waves were made of.

In 1926, Max Born, a brilliant German physicist, had an

interesting idea about what the ψ waves really were. His idea was that they were *waves of chance*.

Waves of chance? Just what did that mean?

Consider these waves of chance in comparison with other kinds of waves. Waves in the ocean could be called "height waves" because it is the height of the water that is different in different locations. Waves sent along a rubber hose would also be waves of height, unless you made the hose ripple sideways, and then the waves could be called "left–right" waves. However, there are waves that are neither height waves nor left–right waves. Sound waves, for example, are "waves of pressure." The sound waves are made up of ripples of high pressure and low pressure in the air.

On the other hand, there are things that are hard to imagine as waves at all. It is difficult, in physics, to imagine waves of friendship, or waves of squareness, or waves of confusion, or waves of truth. These don't seem like the kinds of physical things that would have to do with wave motion. Chance seems to belong in this group of unphysical things.

But the truth is the opposite. There are waves of chance. This is a very hard thing to imagine. But the ripples that move along in waves of chance are made up of places where there is a good chance of finding a particle and places where there is not a very good chance of finding a particle.

The fact is that a wave is *anything* that ripples along in such a way that it goes through rhythmic changes. It doesn't matter what the thing is that is changing; the idea is the same whether or not it is height, pressure, or chance.

If you have a ring of a hundred people wearing hats, and

each person, in turn, takes off his or her hat and then puts it on again, you would have a wave of "hats-condition" that would ripple around the group: "hats-off" places rippling after "hats-on" places all around the ring.

In particle physics, then, the matter waves that make up particles actually are waves of chance. High-chance places ripple after low-chance places. The waves of chance ripple around in circles when the particle is an electron in an atomic orbit. They ripple back and forth when the electron orbit goes straight through the nucleus. They ripple along in straight lines when a free particle is moving through interatomic space.

Suppose that we are dealing with the matter waves that make up an electron. Then a wave peak is a place where the chance of finding an electron is very good. But a low part of the wave is a place where the chance of finding the electron is not very good. The ψ-stuff is chance itself.

Furthermore, when a wave packet is formed, it is possible to know the location of the electron very precisely. That is because the ψ-stuff has bunched up very compactly into a "particle." The existence of a wave packet means that a particle has just been detected. The chances are certain, at that moment, of finding the electron in one certain place, for instance at the point where it collided with another particle.

So far, so good. But what about the fact that the wave packet fades out so fast? Doesn't that still leave us in trouble? Not at all, said Max Born. *After* you detect the free electron (as a particle in a collision, for example), you immediately begin to lose track of where it is. As time goes by, it could go practically anywhere. Its chance of being in one place is about the same as its chance of being in another place. The compact little blob of chance *does* spread out after the observation. This spreading out and flattening of the wave packet represents the observer's loss of information as to where the electron is. It could be anywhere.

Born's ideas about chance have never been outdated. Even now, waves and blobs of chance are the central idea in physics. The electron, the photon, and all other subatomic "particles" can be understood on this basis.

A similar idea was contributed a year later by a nineteen-year-old German scientist who divided his recreation time between physics and skiing. This young scientist's name was Werner Heisenberg, and he had another idea regarding the way we experience subatomic events. He saw that in trying to get one kind of information, we automatically lose other kinds of information.

For instance, consider a collision between a photon and an electron, known as the *photoelectric effect*. A photon of light hits an electron, knocking it loose from the atom and the surface that it was part of. At the moment of impact, you can locate the photon exactly. It was right there, where the electron spurted off the surface and left a track in a cloud chamber. But this experiment loses all information regarding the direction and speed and wavelength of the light.

On the other hand, an experiment such as the double-slit experiment, which provides exact information concerning the motion of the moving wave front, and the wavelengths of the light waves in that wave front, tells us nothing about where any light photon is located.

Even with perfect instruments, the scientist would be faced with these fundamental uncertainties about subatomic events. Heisenberg called this idea the *uncertainty principle*.

Now let's consider a single experiment that sums up all of these ideas—wave mechanics, chance events, and the uncertainty principle—all of which are part of the theory known as *quantum mechanics*. Let's send some electrons through a pair of slits. And let's arrange for these electrons to hit a sensitive film after they go through the slits, leaving bright spots where they hit the film. Suppose that the electrons are coming from a weak radioactive source which ejects only one electron every few seconds. And let's watch the impacts as the electrons hit the sensitive film.

The first electron hits and makes a dot on the film. There doesn't seem to be any question of its particle nature. It hits at a single point, rather than creating an interference pattern. It may not be obvious which slit it went through, for the edges of the slits can scatter particles out of their original path.

As time goes on, there are other particle impacts. Then, gradually, we see a pattern starting to form. The dots from all the individual impacts are forming an interference pattern. Looking at this pattern, it seems clear that waves were traveling through the two slits. But if waves went through the slits, why was each impact clearly a particle impact?

In the language of quantum mechanics, what happened was something like this:

1. A single particle was ejected from a radioactive atom.
2. The particle spread out in waves of chance as it approached the two slits.

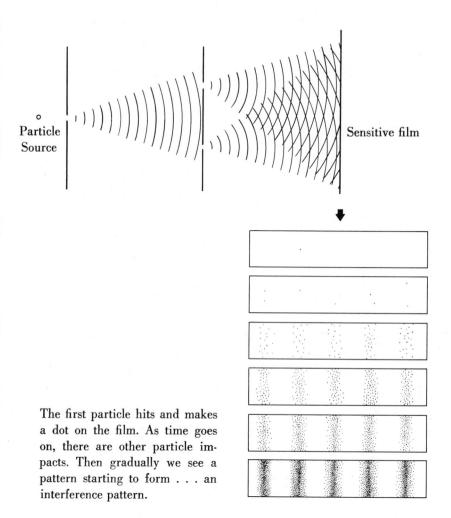

Particle
Source

Sensitive film

The first particle hits and makes a dot on the film. As time goes on, there are other particle impacts. Then gradually we see a pattern starting to form . . . an interference pattern.

Figure 15

Quantum experiment

3. The waves went through both slits.

4. The waves of chance formed an invisible interference pattern, as they arrived at the film. The peaks of this pattern were places where it was very likely that the particle was present. The low parts of the pattern were places where it was very unlikely that the particle was present.

5. The energy of the advancing wave front was transferred to an atom in the film. This meant that the energy had to reappear in the form of a particle. That particle probably hit an atom near one of the higher peaks of the interference pattern.

6. The impact caused a chemical change in the material of the film, resulting in a small dot on the film.

7. The same process took place for the second particle. It too spread out in waves of chance, which formed an interference pattern. It too reappeared as a particle, probably near a peak of the pattern, as it transferred its energy to an atom of the film.

The single experiment described above illustrates several important ideas of quantum mechanics. The moment of collision brought out the particle-nature of each electron in this experiment. This is generally true of subatomic collisions. We can pinpoint the place and time of the collision with great accuracy. But as the particle bounces away from the collision, its speed and direction are completely lost to us.

The particle that we've located so exactly is a wave packet, and it spreads out in all directions after the collision.

You can think of the particle hitting the screen as being

like a pebble dropped in a pond. After the pebble hits the water, ripples spread out across the pond. But in the case of a subatomic particle, it doesn't just *cause* the ripples, it *becomes* the ripples.

The particle was like a little "blob of chance" at the moment of collision. Its chance-nature appears in the fact that, instead of bouncing away from the collision in a definite direction and at a definite speed, it spreads out in all directions at many different speeds as a whole family of waves. Any point in this expanding wave front has a chance of representing the motion of the particle. So, the actual speed and direction of the particle are lost in the collision that located that particle so exactly.

This illustrates Heisenberg's uncertainty principle. Pinpointing the location of an electron at any given moment means losing information about its motion.

On the other hand, what about the waves whose presence became visible in the final pattern of bright and dark bands? From this pattern we learn the wavelengths and energy of the waves. We also know their exact speed and direction of motion, for the motions of a wave are neatly defined by its wavelength and energy. But times and locations relating to the waves are lost. The waves were spread through the entire space between the radioactive source and the screen where the pattern appeared. And they were in motion throughout the entire time interval of the experiment. This too illustrates Heisenberg's principle. Getting one form of subatomic information means losing that information in some other form.

ATOMIC STRUCTURE

10 BEFORE going farther it might be a good idea if we stop and get our bearings by considering atoms. Because they are such well-behaved objects and because of the practical part they play in the structure of matter, atoms are like peaceful harbors where we can rest and get some feeling of where we are, before plunging again into the wild ocean of fundamental particles.

All the substances that we are familiar with are made of

atoms. To begin with, it is useful to think of atoms as small spheres, like miniature tennis balls packed together in huge numbers.

In a *solid*, the atoms are tightly packed in definite arrangements, like soldiers in a parade. By heating a solid, you are adding energy to the atoms so that they start to move around more freely. When this free motion reaches a certain point, the atoms lose their definite arrangement and start to move about as if the parading soldiers broke ranks and started to run in all directions. This change in a substance is called "melting." It is the process by which a solid turns into a *liquid*.

If we heat the liquid, the atoms not only move freely in all directions, they go one step farther and fly apart from one another. It's as if the soldiers who were running in all directions were to sprout wings and go flapping into the sky, each one in a different direction. This change in a substance is called "boiling." It is the process by which a liquid turns into a *gas*.

Every substance can be found in three states: solid, liquid, or gas. In the case of water, for example, we see it as a solid (ice), as a liquid when it melts, and as a gas after it boils away as steam. Actually, steam is the way water looks just as it changes from liquid to gas. After water becomes a gas, it is invisible and becomes part of the air. This gaseous water can turn back into a liquid, which is what happens when a window or the windshield of a car gets misty.

Not every substance is easy to turn into all these forms. It takes extremely high temperatures to turn some things, such as iron, into a gas, and extremely low temperatures to turn some things, such as air, into solids. That's obvious when you stop and realize that you don't hear much about "iron gas" or solid pieces of air. But the basic principle still holds true—with enough effort, and enough high-powered scientific equipment, any substance can be turned into solid, liquid, or gas.

Besides observing which of these states a substance is in, we can look at the kinds of atoms it is made of. If the atoms are all alike, then that material is a "simple substance" or an *element*. Some examples of elements are iron, carbon, oxygen, and sulfur. A solid piece of iron, for example, is made up of a huge number of iron atoms, all alike, held together in a regular formation by interatomic forces. Like the wall on p. 2, this lump of iron is mostly empty space between the atoms. Only the force field within the iron atoms creates the hard, smooth object, the lump of iron that you can feel and hold in your hand.

Many other substances are *compounds*, made up of different kinds of atoms that are joined together. Water, table salt, rust, and sugar are examples of compounds. Water, for instance, is made up of hydrogen and oxygen atoms. Each oxygen atom in the water has two hydrogen atoms joined to it.

A "hookup" like that, in which atoms are joined together, is called a *molecule*. So, a sample of water is made out of molecules, each molecule consisting of three atoms—an oxygen atom and two hydrogen atoms that are "pegged" into it. The molecules of a compound act very much like atoms, moving around at different speeds, depending on how much heat energy they have.

In 1869, a Russian chemist named Dimitri Ivanovich Mendeleyev was trying to understand the properties of the different chemical elements. Trying to find some pattern in the elements, Mendeleyev arranged them by their weights. Beginning with the lightest, least dense element, hydrogen, Mendeleyev called it atomic number 1, and he numbered the other elements in order of increasing density:

Atomic Number	Element	Symbol
1	hydrogen	H
2	helium	He
3	lithium	Li
4	beryllium	Be
5	boron	B
6	carbon	C
7	nitrogen	N
8	oxygen	O
9	fluorine	F
10	neon	Ne
11	sodium	Na
12	magnesium	Mg
13	aluminum	Al
14	silicon	Si
15	phosphorus	P
16	sulfur	S
17	chlorine	Cl
18	argon	A
19	potassium	K
20	calcium	Ca

Figure 16

The list could be continued, since there are 92 elements found in nature, although not all of these were known in Mendeleyev's time. In fact, chemical elements heavier than any found in nature have been generated in recent years through particle bombardments in which the bombarding particles were absorbed into atomic nuclei. As a result, there are now 103 known elements.

Here is how Mendeleyev arranged the first 20 elements, in order of their atomic numbers:

1 H							2 He
3 Li	4 Be	5 B	6 C	7 N	8 O	9 F	10 Ne
11 Na	12 Mg	13 Al	14 Si	15 P	16 S	17 Cl	18 A
19 K	20 Ca						

Figure 17

Mendeleyev saw that the elements in the same vertical column of this table are alike in certain ways. For instance, atomic numbers 2, 10, and 18 (helium, neon, and argon) are all colorless gases that are *inert*—that is, they do not combine chemically with other elements. So these three elements, and other elements of greater atomic numbers in that same column of the table, are known as the inert gases.

On the other hand, fluorine (atomic number 9), chlorine (atomic number 17), and other substances in that column are greenish or brownish gases or liquids that are very active chemically. That whole group is known as the *halogens*. Beryllium (atomic number 4), magnesium (atomic number 12), and calcium (atomic number 20) are rather soft, white metals, known as the *alkali earths*.

In patterns of this kind, the properties of the various elements repeat periodically, so that definite *families* or groups of elements can be found. The complete table that lists all of the elements in this way is called the *periodic table of the elements*.

Mendeleyev did not offer any reason for the pattern in the properties of the elements. His finding was temporarily accepted by scientists as an unexplained fact of nature.

But twentieth-century research into atomic structure, arising out of the work of physicists such as Rutherford, Bohr, and Pauli, showed what lies beneath the periodic table. The repeating properties of the different elements relate to the orbits of electrons.

Figure 18 does not show the different Bohr orbits that each electron can be in. Instead, each electron is shown in a circular orbit. The emphasis here is not on the motions of individual electrons from orbit to orbit, but on groupings of electrons in *orbital shells*.

For hydrogen, we see the single electron in orbit. For helium, two electrons are in orbit. It happens that this first shell is completely filled by two electrons. So, when we look at atomic number 3, lithium, the third electron has to be located in the next shell. Therefore, lithium has an inner shell with two electrons and an outer shell with a single electron.

Notice that the atomic number for each element is equal to the number of electrons in orbit. Hydrogen, atomic number 1, has one electron; helium, atomic number 2, has two electrons; lithium, atomic number 3, has three electrons, and so forth.

The first shell, as we have seen, is filled up by two electrons. The second shell, as can be seen in Fig. 18, can hold eight

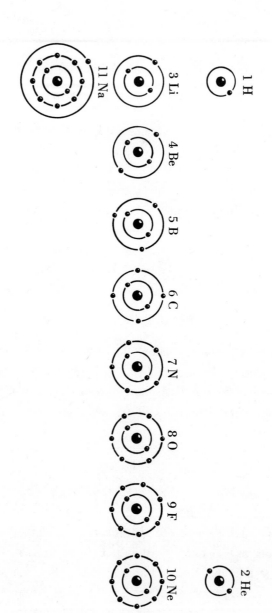

The arrangement of electron orbits of the first eleven elements

Figure 18

· · · 70

electrons. Consider atomic number 10—neon. It can hold its ten electrons in two shells, with two electrons in the first and eight in the second shell. But sodium, atomic number 11, needs a third shell for its eleventh electron.

Now it becomes clear why there are periodic repetitions in the properties of elements. Helium and neon, members of the same group, both have complete outer shells—helium with its two-electron first shell, and neon with its eight-electron second shell. And the other inert gases, such as argon, carry out this theme. All of the inert gases have full outer shells with no electrons left over.

Hydrogen, lithium, and sodium, on the other hand, each have a single electron in the outermost shell. That is what makes them alike. In general, elements are alike in their properties if their outermost electron shells are alike.

The chemical behavior of an element is based on the fact that each atom "tries" to have a complete outer shell. Atoms combine chemically in ways that give each atom a complete outer shell. For instance, two atoms of hydrogen join as shown in Fig. 19a to form a hydrogen molecule.

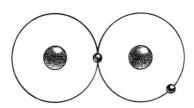

Figure 19a

Two atoms of hydrogen join like this to form a hydrogen molecule

The two electrons are shared by both of the hydrogen atoms. Therefore, each atom "feels" as if it has a complete shell with two electrons in it.

Two atoms of hydrogen (H) join with one atom of oxygen (O) to form a water molecule (H_2O). The situation is as pictured in Fig. 19b.

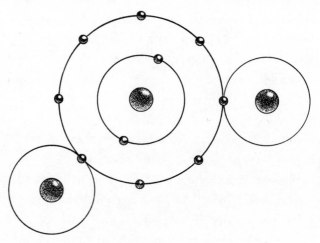

Figure 19b

Two atoms of hydrogen join with one atom of oxygen to form a water molecule

The chemical behavior of an element is based on the fact that each atom "tries" to have a complete outer shell

Each hydrogen atom feels complete because it has "given up" its electron and, therefore, feels as if it has no outer shell. The oxygen atom feels complete because its six-electron outer shell has now become a full eight-electron outer shell.

An atom of sodium gives its single outer electron to an atom of chlorine to form sodium chloride, or table salt. In this case, the electron is not just shared but is totally given away. So a sodium ion and a chloride ion are formed, which drift away on their own in a solution of salt water. They both are electrically charged—the sodium ion being positively charged from having lost one negatively charged electron, and the chloride ion being negatively charged from having gained the electron.

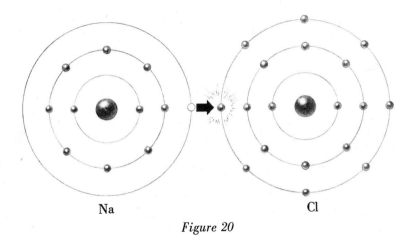

Na   Cl

Figure 20

A table-salt molecule forming

Salt water is a loose swarm of the two kinds of ions, moving freely about. Dry table salt is a crystal structure of these two kinds of ions held together by electrical forces. It is interesting to notice how different table salt is from the two elements that make it up. Sodium is a white metal that reacts violently, almost explosively, when it is dipped in water. Chlorine is a heavy greenish-yellow poisonous gas. Yet, when they combine to complete their outer shells, they turn into something far different from either of these things.

The most chemically active elements are those such as sodium that have only a single outer electron to give away, and those such as chlorine that need only one more electron to complete their outer shells. Atoms such as boron, carbon, or nitrogen, which have to give or receive several electrons to complete their outer shells, do not react so easily.

Finally, there are atoms such as helium, neon, and argon, which start off with complete outer shells. Such atoms do not react with other atoms at all. As a result, these elements are inert.

The number of electrons needed to complete a shell is a result of Pauli's exclusion principle. Each electron in the shell has to have a different set of quantum numbers from every other electron in the shell. When all possible sets of quantum numbers are used up for a particular shell, the shell is complete. Any additional electron goes into a new shell. A new shell means that the principal quantum number, n, has changed, and therefore the exclusion principle has been satisfied.

So, the electron shell structure provided a logical basis for Mendeleyev's periodic table, and it explained the chemical behavior of the elements. But what about the nucleus, what about its structure?

THE NUCLEUS

11 A hydrogen atom has one negatively charged electron. In the language of quantum mechanics, the charge of an electron is −1. Therefore, in order for the electric charge to be balanced, the nucleus must have a positive charge of +1. A helium atom has two negatively charged electrons, so its nucleus must have a charge of +2. In going through the periodic table, we find, then, that the nucleus of each element has a positive charge equal to the atomic number of the element. But, while the charge on the nucleus just balances the total charge of the orbital electrons, the mass of the nucleus is far greater than the mass of all the electrons.

When particle bombardments broke the nuclei of atoms apart, two heavy particles were discovered to be the main ingredients of the nucleus. One of them was a positively charged particle having a charge equal and opposite to that of an electron. This particle, discovered by Rutherford in 1919, was named the *proton*. Its mass was determined to be about 1,836 times as great as the mass of an electron.

In 1932, the English physicist James Chadwick discovered a second nuclear particle, slightly heavier than the proton, with a mass of 1,840 electrons. This particle was named the *neutron*. It is electrically neutral.

Physicists have a convenient name for the heavy nuclear particles. They call them *nucleons*. A nucleon can be either a proton or a neutron. In fact, except for electric charge, the proton and the neutron are identical for all practical purposes. It is useful to think of them as two different forms of the more general particle, the nucleon.

The nucleus of the hydrogen atom is made up of a single proton. The helium nucleus, as atomic number 2, must have two protons in its nucleus. But the helium nucleus is not just twice as heavy as the hydrogen nucleus—it is four times as heavy. Physicists concluded from this that it must have two neutrons as well as two protons.

This pattern continues throughout the periodic table. That is, the number of protons in a nucleus always equals the atomic number of the element. But, except for hydrogen, every nucleus

Hydrogen nucleus

Helium nucleus

P+

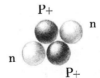

Figure 21

has neutrons. The neutrons contribute mass, but do not add to the electric charge of the nucleus.

Actually, the hydrogen nucleus can also have neutrons. There are three different kinds of hydrogen, having nuclei like this:

Hydrogen

Atomic number 1	P+ Hydrogen 1	Atomic weight 1
Atomic number 1	P+ Hydrogen 2 n	Atomic weight 2
Atomic number 1	P+ Hydrogen 3 n n	Atomic weight 3

Figure 22

All three of these nuclei are hydrogen nuclei, because there is a single proton in the nucleus, which is what defines hydrogen as atomic number 1. But the three kinds of hydrogen each have different atomic weights. The *atomic weight* of any atom is approximately equal to the total number of heavy particles in the nucleus—both protons and neutrons. The *atomic number* refers only to the number of protons in an atom.

In any sample of hydrogen, most of the atoms are hydrogen 1. A very small fraction of the atoms are hydrogen 2 and an even smaller fraction are hydrogen 3. When an atom of some particular atomic number is found in different atomic weights, it is said to have different *isotopes*. Hydrogen 1, hydrogen 2, and hydrogen 3 are all isotopes of hydrogen.

In general, atoms tend to have numerous isotopes—some occurring in nature, and many others existing only as a result of artificial transformations under nuclear bombardment. For instance, atomic number 6, carbon, has ten different isotopes. The most common of these is carbon 12, with a nucleus of six protons and six neutrons.

Carbon 14, with six protons and eight neutrons, is another important isotope of carbon. It is a radioactive isotope, giving off a low intensity beta radiation. The process, occurring within the nucleus, is this: A neutron emits an electron, and becomes a proton. The electron that is emitted was not an electron in orbit around the nucleus. Rather, it is an electron that came into existence at the moment that the neutron hurled it out, just as a shout comes into existence at the moment it comes out of a person's mouth.

After this event, the carbon nucleus has gained a proton

and lost a neutron; it now has seven protons and seven neutrons. Its atomic weight is still the same as before (14), but now it has an atomic number of 7. That means that it has become a nitrogen atom—nitrogen 14.

Radioactive isotopes are much more common among the heavier elements. For instance, uranium has an atomic number of 92. It has 92 orbital electrons and 92 protons in its nucleus. The most common isotope of uranium is uranium 238. This is the substance the French physicist Henri Becquerel was working with in 1896 when he discovered radioactivity. The number of neutrons is $238 - 92$, or 146. This is an unstable isotope, which breaks down by ejecting alpha particles. The breakdown generates other unstable isotopes, which break down in turn, emitting beta particles (electrons) and gamma rays (high-energy photons).

Radioactive breakdowns of this kind, from various heavy unstable isotopes, provide a radiation of various particles, and with these particles, provide clues to the structure of the atomic nucleus.

THE NEUTRINO

12 SOMETIMES a neutron breaks apart into a proton and an electron. This process is shown in the equation:

$$n \quad \rightarrow \quad p^+ \quad + \quad e^-$$
neutron   proton   electron

When a neutron breaks apart inside the nucleus of an atom, it ejects an electron, which shoots out of the nucleus in the form of beta radiation. A passing electron is captured, to go into orbit around the nucleus, balancing the electrical charge of the new photon. The rest of the neutron stays in the nucleus as a proton. The original atom has changed into a different kind of atom (see the carbon 14 example on p. 78).

In 1950, physicists found that a free neutron, existing outside the atomic nucleus, does not last forever. On the average, it has a lifetime of about twelve minutes. Then it breaks apart into a proton and an electron, as shown in the equation on p. 80. When the neutron is inside a nucleus with protons and other neutrons, it usually does not break apart, unless the nucleus happens to be an unstable, radioactive one.

But even before 1950, a problem arose concerning the process of neutron breakdown. Some energy was lost in the process, and nobody knew where it went.

What happened was this. An atomic nucleus would disintegrate through beta decay, ejecting an electron. When the mass of the electron and the mass of the new, lighter nucleus were added together, they would not equal the mass of the original nucleus. Some mass was lost. By Einstein's principle of mass-energy conversion (see p. 51), it was logical to think that the lost mass was converted into the energy of motion of the ejected electron. In some cases the ejected electron had enough energy to account for almost all of the lost mass. In other cases, however, the electron had very little energy, and most of the lost mass was not accounted for. In general some loss in mass-energy was taking place. But where was it going?

That was not the only problem. There was also a question of "recoil." When a gun fires a bullet, there is a backward "kick" of the gun. This kick is what is meant by recoil. When a nucleus fires an electron, there is also a recoil. That is, as the electron goes off at enormous speed in one direction, the nucleus should move slightly in the opposite direction. There is an exact law of Newtonian mechanics that describes how this should happen—the ejection speed of the electron times the mass of the electron should equal the recoil speed of the nucleus in the opposite direction times the mass of the nucleus. But, in general, this did not happen. There was a serious imbalance. Something had to be absorbing the lost part of the recoil.

The best models in physics are the ones that explain observed events with the fewest possible assumptions. That is, a good model is compact. Therefore it made sense to suppose that some unknown particle was carrying away the missing energy and also absorbing the extra recoil. This new particle had to be something that was very hard to detect.

In 1931, Wolfgang Pauli predicted the existence of this new particle, the *neutrino*. Pauli was a physicist very much like J. J. Thomson. That is, he wasn't at all good at experimental work and rarely went near the scientific laboratory. In fact, he didn't even write equations very often, nor work with pads, nor chalkboards. Instead, most of his work hours were spent sitting in a chair or going for walks and just thinking.

Pauli was able to predict what the new particle would have to be like.

$$n \quad \rightarrow \quad p^+ \quad + \quad e^- \quad + \quad \nu$$

neutron   proton   electron   neutrino

Physical reactions of this kind have to be able to go in reverse. The reverse process may be extremely unlikely, but it has to be theoretically possible. The reverse reaction would be one in which the proton, electron, and neutrino recombine to form the neutron again.

$$p^+ \quad + \quad e^- \quad + \quad \nu \quad \rightarrow \quad n$$
proton   electron   neutrino   neutron

In this reverse reaction, the equal and opposite charges on the proton and the electron must cancel out in forming the electrically neutral neutron. Therefore, the neutrino has to have zero charge.

Pauli also concluded that the neutrino would have to have spin. In Chapter 6, we met some quantum numbers having to do with the condition of an electron in an atom. The quantum numbers n, l, and m all have to do with the electron orbit, its size, shape, and tilt. But the fourth quantum number, s, describing the spin of the electron, is a quantum number that describes an electron whether or not it is in orbit around a nucleus. Every electron has ½ quantum of spin. Nucleons also have ½ quantum of spin. A photon, on the other hand, has 1 quantum of spin. Later, other particles are described, having various amounts of spin. However, in quantum mechanical terms, that spin is always a half or a whole quantum: 0, ½, 1, 1½, 2, etc. The energy of the spin is never such that it lies in between these quantum levels, such as ¼, 1⅓, or anything of that sort.

The spins of the neutron, proton, and electron are such that there is (along with the missing energy and recoil) a missing ½ quantum of spin. The neutrino would have to account for the lost spin.

Further analysis of this peculiar particle showed that spin is the only thing that it has. Not only does it have zero charge, it also has zero mass. The neutrino is a spinning little bit of nothingness that travels at the speed of light. That is as fast as anything can possibly go. Einstein had proved in 1904, in his special theory of relativity, that the speed of light is an upper limit in the universe. Nothing can ever travel faster than light.

So far, the neutrino was undetected. Its existence was strictly theoretical. It *had* to exist to account for the missing energy, recoil, and spin. And it had better not exist very much, because it could have neither mass nor charge.

Physicists were a little uneasy with this weird particle. The whole neutrino idea was open to an obvious criticism; it looked like a bit of circular reasoning designed to cover up losses that were simply not understood.

They became even more uneasy when calculations showed that it would be practically impossible ever to detect a neutrino. Most particles can be absorbed by a few centimeters or a few meters of matter. That means that they can be absorbed in a detector such as a cloud chamber or a Geiger counter. But the calculations showed that a neutrino, on the average, would penetrate 3,500 light years of solid lead before interacting with matter. A light year is the distance that light travels in an entire year, moving at its enormous speed of 186,000 miles every second. A stretch of 3,500 light years of solid lead is total fantasy. Of course no such stretch of matter exists in the universe. But the calculation was no fantasy, and indicated that neutrinos would probably never be detected and that they would have to remain doubtful and mysterious forever.

However, a way was eventually found to capture the neutrino, and the *antineutrino*. (The antineutrino is a neutrino with a backward spin.) The antineutrino is important because later calculations showed that neutron breakdown involves the antineutrino, not the neutrino. So, the correct formula is:

$$n \quad \rightarrow \quad p^+ \quad + \quad e^- \quad + \quad \bar{\nu}$$

neutron   proton   electron   antineutrino

Antineutrinos are just as hard to detect as neutrinos, if not harder. They too would (on the average) penetrate 3,500 light years of lead before interacting in any way with matter. How do you detect something like this in a few centimeters of a radiation detector?

The answer is to have an enormous number of antineutrinos going through the detector. The antineutrino that would penetrate 3,500 light years of lead is just the average antineutrino. But this penetration is a chance event. Other antineutrinos would penetrate farther than that. Others would not penetrate as far. Once in a very great while, one would even interact with matter in a radiation detector. If detectors could be put near a source that produced enormous numbers of antineutrinos, this kind of event would happen fairly often, and antineutrinos would then be detectable.

The flow of antineutrinos needed was really extreme. If a

million antineutrinos went through the detector every second, only one of them would be detected in a million years.

Fortunately a kind of source is available that can send far more than a million antineutrinos per second through a detector. A nuclear-fission reactor can provide that kind of antineutrino flow. Two American physicists, Clyde Cowan and Frederick Reines, set up huge detecting tanks near a reactor at Savannah River, South Carolina, in 1953. According to their calculations, this reactor would pour out a million antineutrinos every billionth of a second. In this situation, using some very clever techniques, the two physicists detected antineutrinos. The mysterious particle had finally been trapped.

All natural radioactive isotopes and all atomic reactors emit antineutrinos, not neutrinos. But just the opposite is true of the sun and the other stars. Stars get their energy by nuclear reactions that happen to give off neutrinos. So, space in general should be flooded with neutrinos. They are even harder to trap than antineutrinos because physicists cannot conveniently put detectors next to the sun the way they can put them next to an atomic reactor.

To detect neutrinos, physicists put detection tanks a mile deep in an abandoned silver mine. This keeps the tanks away from cosmic radiation and other kinds of radiation that would interfere with the delicate experiment. For neutrinos, a mile of solid rock is no obstacle at all. A neutrino could go through the entire earth with complete ease. To a particle capable of penetrating 3,500 light years of lead, the earth is like a piece of fluff.

The sun's gigantic flood of neutrinos is considerably spread out by the time it reaches the earth, so the earth gets only a small portion of that flood. Nevertheless, in 1965, seven neutrinos had been detected over a nine-month period of observation.

So, these strange spinning bits of nothing are real and can be captured. Wispy though they are, they play an important part in the activity of the universe.

It's strange to realize how many of these particles are darting past without interacting with anything. The noted scientist and author, Isaac Asimov, calculates that about 2,000 antineutrinos go through a person's body every second. Yet, the chances are only one in a billion that one of these antineutrinos will interact with a proton in that person's body during a seventy-year lifetime. Another way to think of it is that there is probably only one person alive today who has absorbed an antineutrino.

VIRTUAL PARTICLES

13 AS soon as scientists realized that the atomic nucleus was made up of protons and neutrons, it became obvious that a strange new force was operating in the universe. The laws of electromagnetism clearly showed that the protons, having positive electrical charges, should push each other apart, and the nucleus should therefore burst outward. The force of gravity within the nucleus would not be enough to keep them together, and the uncharged neutrons wouldn't have any effect upon the electrical situation.

Here was another of those strange catastrophes, involving

the collapse of matter, which wasn't actually happening. Why wasn't it happening?

The answer had to be that a powerful attractive force must be at work among the protons and neutrons in the nucleus. This force would be so powerful within the nucleus that it would overcome the electromagnetic forces. But this strong force would have to vanish very quickly outside the nucleus, because there is no trace of it to be seen in the interactions between the nucleus and the orbital electrons. Also, it has no effect on forces between neighboring atoms. So a strong nuclear force has to be extremely powerful at very close ranges, but it has to fade out at longer ranges.

How short is the range of the strong force? To describe that range, it is convenient to bring in a new unit of length, because the angstrom unit is too big. The new unit is called a *fermi* (F). It takes 100,000 fermis to make 1 angstrom unit. The distance between two neighboring nucleons is about 1½ fermis. The strong force must be able to operate across that 1½-fermi range, up to about 4 fermis but no farther.

Something strange seems to take place between two neighboring nucleons as they are pulled together by the strong force. To understand what is going on, perhaps it would be helpful to take another look at electromagnetic forces.

In 1932, Werner Heisenberg concluded that charged particles bounce photons of light back and forth between them. This exchange of photons is the way that electromagnetic forces act between the particles.

For example, the hydrogen atom has a proton for its nucleus and one electron in orbit. Electromagnetic energy keeps the electron in orbit around the proton. According to Heisenberg, the proton shoots a photon at the electron. The electron shoots the photon back at the proton. This rapid-fire exchange goes on all the time. In fact, the proton and the electron keep many photons going back and forth at one time. In his book, *The Strange Story of the Quantum*, Banesh Hoffman compares protons and electrons to fantastic tennis players, able to rally with a dozen balls in motion at once.

Heisenberg was able to show that this exchange of photons kept the two oppositely charged particles attracted toward each other. He was also able to show how the exchange of photons would drive apart two particles of like charge.

So, according to Heisenberg, every charged particle has a little swarm of photons darting in and out of it like bees going in and out of a beehive. He called these photons *virtual photons*. By saying that the photons were "virtual," he meant that they weren't exactly "real."

Let's explore this idea further, to see what Heisenberg meant by virtual photons and real photons. Real photons are any photons that can be detected—for instance, the photons of sunlight or electric light that are lighting this page are real. The fact that you are seeing these words means that you are detecting the photons that reflect from the page.

The virtual photons that travel back and forth between charged particles are not real in this way. They are not detectable; that is, they do not light pages or leave images on photographic plates or eject photoelectrons from surfaces. They exist only in the strange hidden game between the charged particles.

However, if enough additional energy is suddenly put into the charged particle, one or more of its virtual photons becomes real and can then do all the things that real photons do. This result fits the fact that charged particles, when accelerated, do radiate light. Using Hoffman's tennis idea, it's as if one of the players suddenly took a wild swing and hit a tennis ball out of the court and over the fence.

Forces between two particles involving this exchange of virtual photons were named *exchange forces*. Of course, these exchange forces were not a new kind of force. They were just a new way of describing electromagnetic forces.

The Japanese physicist Hideki Yukawa felt that exchange forces might also be used to describe the strong force between nucleons. But what kind of virtual particles would the nucleons exchange in their tennis game? Virtual photons? No, some different kind of virtual particle was needed.

Yukawa's calculations showed that a very high-energy particle was needed to hold the nucleons together. But where would the energy come from to allow this particle to exist at all, even as a virtual particle? Yukawa's logic kept leading to the same result—that a nucleon would not have the energy to produce one of these virtual particles. Some additional energy definitely was needed.

Yukawa found an unusual way out of this problem. He knew how far his virtual particles had to travel—just 1½ fermis, the distance between neighboring nucleons. In fact, they couldn't travel much farther than that because if they did the strong force would be felt at greater distances, outside the nucleus, which it is not. Yukawa also had theoretical reasons for assuming that these virtual particles would travel at nearly the speed of light.

Knowing the distance the particles have to travel and

knowing their speed, he immediately calculated how long a virtual particle would have to exist, during its trip between the nucleons. He found out that its lifespan would be very small, even by subatomic standards.

Using Heisenberg's uncertainty principle, Yukawa put these facts together in a very ingenious way. His logic went something like this: This virtual particle exists for an extremely small fraction of a second. It exists for so brief an instant that its "time of existence" is very exact. Because its time of existence is so exactly pinpointed, there is a great uncertainty in the energy of the virtual particle.

In other words, when a certain situation exists for an extremely short time, the uncertainty principle says that there is room for tremendous doubt about the total energy involved in the situation. Because of that doubt, there is energy available that does not have to be accounted for. The uncertainty in the situation allows energy to come from out of "nowhere." In this way, Yukawa's virtual particles could "slip" in and out of existence.

Further calculations showed that each of these particles should be about 250 times as heavy as an electron. So here was a middleweight particle that was in between the lightweight electrons and the heavyweight nucleons. According to Yukawa's prediction, if enough energy were poured into an atomic nucleus, some of these middleweight particles were eventually knocked out of their tennis courts to become real particles that could be detected in cloud chambers and by other instruments. Meanwhile, he was confident that the virtual particles were there, bouncing back and forth between the nucleons, holding the cores of atoms together.

MUONS
AND
PIONS

14 YUKAWA predicted the existence of the middleweight particles in 1935. Within a year a middleweight particle was detected by a young American physicist named Carl D. Anderson. The new particle was only 207 times as massive as an electron, a little lighter than what Yukawa had predicted. But there was room for the difference, since Yukawa's calculation was only approximate. It did seem that this new particle was the Yukawa particle everyone had been waiting for. The particle was called the *meson,* from the Greek word *meso* meaning "middle."

However, there was a much more serious problem with the meson than the fact that its mass was a little low. The problem with the meson was that it could penetrate a few feet of solid lead, shooting past the vicinities of many atomic nuclei, before being absorbed. In other words, it was slow to interact with nucleons. But the Yukawa particle's whole reason for existing was to interact with nucleons. Then, this could not be the Yukawa particle, which was meant to be the tennis ball used by the nucleons and had to be very easily absorbed. Anderson's meson had made things more puzzling, not less. If it wasn't the Yukawa particle, then what was it? And how did it fit into the subatomic picture?

A year later, English physicist Cecil F. Powell detected a somewhat heavier meson. This meson was 270 times as massive as the electron and interacted very strongly with atomic nuclei. Everyone was sure that this was the Yukawa particle.

To distinguish these two kinds of particles, physicists named this heavier one the "pi-meson" and named the lighter one the "mu-meson." In time, these were shortened to *pion* and *muon.* The following statements sum up the main characteristics of pions and muons:

1. The pion is the Yukawa particle. As virtual particles, pions are the agents of the strong force between neighboring nucleons.

2. Pions are 270 times as massive as electrons. There are three kinds of pions: positively charged, neutral, and negatively charged.

3.   A pion has an average lifetime of about one hundred-millionth of a second. When a charged pion decays, it usually produces a muon and a neutrino or antineutrino.

4.   The neutrinos (and antineutrinos) that are produced along with muons are not exactly the same as the neutrinos (and antineutrinos) that are produced along with electrons during beta decay. The muon neutrino is very much like the electron neutrino in being a massless, uncharged, spinning little bit of energy, capable of penetrating 3,500 light years of lead. But muon neutrinos react slightly differently with other particles compared with the way that electron neutrinos react, and they definitely have to be considered as different particles.

5.   There are positively charged muons and negatively charged muons, but no neutral muons. Muons decay into electrons and neutrinos (or antineutrinos). The average lifetime of a muon is about one millionth of a second.

In general, the pion fits very neatly into the world of particles, providing as it does the binding force between nucleons. But its strange daughter, the muon, doesn't fit at all. It seems like an "unnecessary" particle. The fact that muons result from the breakdown of pions does relate the muon to other particles. But that relationship is not enough to give any real significance to the muon. It is still unnecessary.

What do physicists have in mind in thinking of some particles as necessary and others as unnecessary? A necessary particle is one that has to do with how the universe is put together. Protons and neutrons are necessary because they form the nuclei of atoms. Pions are necessary because they bind these nuclear particles together. Electrons are necessary because they orbit or ripple around the atomic nuclei to complete the atomic structure. Photons are necessary because they provide the electromagnetic force that keeps the electrons in orbit. Electron neutrinos are necessary because they absorb the recoil, energy, and spin of various nuclear events.

But the muons (and the muon neutrinos that accompany them) seem to play no part in the structure of things. It seems as if the universe could get along quite well without them.

The negative muon acts very much like a gigantic electron, and the positive muon acts like a gigantic proton. A muon also resembles an electron in having ½ quantum of spin. Physicists have been able to observe situations in which a negative muon briefly takes the place of an electron in an atom. For the brief lifetime of the muon, for one millionth of a second, it acts like an electron in a Bohr orbit.

Actually, one millionth of a second is not so brief on a subatomic time scale. Both the pion with its lifetime of one hundred-millionth of a second and the muon with its lifetime of a millionth of a second are extremely long-lasting. Compare these times with the time that it takes a pion to interact with a nucleon, and they are enormous time spans. Comparing the pion–nucleon interaction to the pion lifetime is like comparing one second to 100 million years. Comparing the pion–nucleon interaction to the muon lifetime is like comparing one second to ten billion years.

This kind of time difference appears in many subatomic events. There are the fast events, such as pion–nucleon interactions. And there are the slow events, such as pion or muon decay. The occurrence of a fast event means that a strong force is operating. The occurrence of a slow event means that a weaker force is operating.

So, it became obvious to physicists that they were dealing with two different nuclear forces, not just one. In addition to the strong force that held nuclei together, there was a far weaker force at work in the universe.

THE FOUR FORCES

15 THERE are four forces in the universe. Every physical process that takes place is a result of one of these forces.

The greatest known force is the *strong nuclear force*. It is often called simply "the strong force." This is the force that holds protons and neutrons together within an atomic nucleus. The release of "atomic energy" involves tapping a very small fraction of the strong force.

The second greatest force in the universe is the *electro-*

magnetic force. It is 137 times weaker than the strong force. This is the force that we have met in earlier chapters in the form of electrical and magnetic forces acting between charged particles.

There is a third force that is far smaller than the first two. This is the *weak nuclear force,* also known for short as "the weak force." It is 100,000 billion times weaker than the strong force. In other words, this weak force is so faint compared to the strong force and the electromagnetic force that it barely exists. But it does make itself known, nevertheless. It is the force that causes electrons and photons to pop out of the nuclei of radioactive atoms. It is the force that Becquerel became aware of when he noticed the radiation given off by a lump of uranium.

The fourth and weakest of the forces is so small that it is only like a shadow in the universe. What is strange is that this insignificant force is the one that looks the biggest to people and seems the most noticeable in our daily lives. It is the *force of gravitation,* the force that makes objects fall toward the ground, keeps satellites and planets in their orbits, and holds the solar system together. This seemingly powerful force is so much smaller than the weak force that ordinary numbers can scarcely explain how the two forces compare. If we let the entire Pacific Ocean represent the size of the weak force, then gravitation would be as big as a single drop of water.

To show just how weak the gravitational force is, consider what happens when electrons and other charged particles hurled out from the sun reach the vicinity of the earth. Such particles would "feel" both the gravitational field of the earth and the magnetic field of the earth.

The earth has a very weak magnetic field. As a magnet,

FORCE FIELDS

The *force-field* idea is found throughout physics and appears in many different models. To say that some portion of space contains a force field means that particles (or larger objects) moving into that space are forced into some new kind of motion. They are speeded up or slowed down or swept into curved paths.

The space around a bar magnet is a good example of a force field. If a bit of iron or steel is placed in the right part of that field, it is pulled into motion and rushes to the end of the magnet. There is no "force" operating near the magnet until the bit of iron is put there. Something has to be pulled before we can say that a force is at work. But the space around the magnet is unusual. It is a space where forces are ready to go into action as soon as something is put there for them to act on. That is what physicists mean by a "force field."

our planet is not at all powerful. For about a dollar, you can buy a magnet small enough to hold in your hand that is much more powerful than the earth's magnetic field. You can prove this by holding a small bar magnet next to a magnetic compass needle. Unless the magnet you buy is fantastically weak, it will overpower the earth's magnetic field. The compass will stop pointing at the north pole of the earth and will point at the "north pole" of the bar magnet.

Yet, even the earth's weak magnetic field is vastly more powerful than the earth's gravitational field. Remember that when we talk about electromagnetic forces in comparison to gravitation, we are going even further than we went in comparing the weak force to gravitation. Now we are comparing 1,000 billion Pacific Oceans to a single droplet of water. The comparison is so huge that it is practically incomprehensible.

The result of this comparison is that the charged particle does not respond to the earth's gravitational pull. That pull is totally lost in comparison to the magnetic force. So the charged particle becomes trapped in a "radiation belt" of particles that spiral around in the earth's magnetic force field. It is not pulled into a gravitational orbit.

It is strange that gravitation, which is not noticeable in the world of the electron, is so large in our lives. It is particularly strange to look at the sheer face of a steep cliff and to realize that this frightening height has to do with a force as feeble in the universe as the beat of a butterfly's wing in a hurricane. (Actually, one beat of a butterfly's wing in a hurricane would be much more significant than the force of gravitation is in comparison with the greater forces.)

The reason that we feel this negligible force is that the

three larger forces are "locked up," in various ways, in the structure of matter. Since most large objects (including the sun, the earth, and our own bodies) are electrically neutral, they do not "feel" the electromagnetic force. As to the strong force and the weak force, for all their power, they act only over an extremely small range. They do not extend outside the nucleus of the atom. This leaves gravitation as the major force at large in the universe, free to act upon stars, planets, spaceships, hailstones, birds' nests, dust specks, people, and all other large objects.

However, the other forces do act on us in less obvious ways. They hold the atoms of our bodies together. They provide the energy of the sun. They are responsible for the solid structure of matter. When you push or bump against an object, you are feeling electromagnetic forces.

There are many examples of particle events that are governed by the weak force. For example, the decay of a pion is always a weak process, which happens in about one forty-millionth of a second.

Not all weak processes take exactly the same amount of time. Various other factors, such as the amount of mass-energy available for a certain reaction, can greatly change the speed of a reaction. In fact, neutron decay, which takes an average of twelve minutes to happen, is an event that proceeds by the weak force, although the time involved is unusually long.

Various physicists, starting with Hideki Yukawa, have predicted the existence of a *W-particle* that acts as the carrier of the weak force, just as pions are carriers of the strong force and as photons are carriers of the electromagnetic force. If the

W-particle exists, then it is formed, along with a proton, at the time of neutron decay, as in Fig. 23. Then, the *W*-particle would decay in turn into an electron and an electron anti-neutrino.

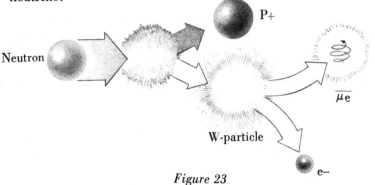

Figure 23

Neutron decay involving an intermediate *W*-particle

Calculations show that the *W*-particle would have to be a massive, very short-lived particle. So it would be very difficult to detect. And, so far, nobody has succeeded in detecting one. The *W*-particle would be so short-lived that an experimenter would only be able to observe the final three fragments of neutron breakdown—the proton, the electron, and (with great effort) the electron antineutrino. The intermediate weak particle would not be observed, and it would seem as if the neutron had broken directly into the three final particles.

Even though the *W*-particle would be quick to decay, its decay would not be quick enough to be a strong event. And, in spite of its somewhat quick decay, it generally would be slow to form, which fits the fact that a weak event is slow to take place.

STRANGENESS

16 THERE are several puzzles, other than the mystery of the *W*-particle, that have arisen in connection with the weak force. One of these has to do with several new particles that have been discovered, starting in the mid-1950s. These include some very large particles, heavier than nucleons—*xi* (Ξ), *lambda* (Λ), *sigma* (Σ), and *omega* (Ω) *particles*. There are also some new mesons, heavier than pions, that were named *k*-particles or *kaons*.

There was a serious problem concerning these new particles. They were all created in strong interactions, but decayed slowly in weak interactions. This was very unexpected. Physicists were used to the idea that particle events are essentially reversible. A particle born in the instant of operation of the strong force should not fade away over the vast ages of time (such as one billionth of a second or even one millionth of a second) required by the weak force. Because of this peculiar imbalance, these new particles were given the name of *strange particles.*

One reaction that produces strange particles is a collision between a negative pion and a proton:

$$\pi^- \quad + \quad p^+ \quad \rightarrow \quad \Lambda^\circ \quad + \quad K^\circ$$

negative pion	proton		lambda	neutral kaon

Figure 24

The collision destroys both the pion and the proton, and two new particles are created, both of them "strange." One is the lambda, a heavy uncharged particle about 1.2 times as massive as a nucleon. The second is a kaon. Kaons can be positively charged, negatively charged, or neutral, and the one formed in this reaction is neutral. A kaon is a meson about $3\frac{1}{2}$ times as heavy as a pion.

The reaction is a strong one. The pion–proton collision and destruction takes place in the usual flicker of time that characterizes strong interactions. But then, the lambda and the kaon last practically forever—a little under one billionth of a second for the average lambda, and anywhere between one ten-

billionth and one millionth of a second for the average uncharged kaon.

Before we look at the decay products of the lambda and
the kaon, let's look again at the earlier equation to observe how
various physical quantities are conserved.

To begin with, the two colliding particles have equal and
opposite charges: the pion with −1 quantum of charge and the
proton with +1 quantum of charge. The newly created particles
are both electrically neutral. So the total charge of the system,
both before and after the reaction, is zero. Charge is conserved.

Spin is also conserved. The pion has 1 quantum of spin,
and the proton has ½ quantum of spin. So the total spin of the
particles entering the reaction is 1½. The kaon has 1 quantum
of spin, and the lambda has ½ quantum of spin. So the total
spin of the particles leaving the reaction is also 1½.

Many other quantities are "conserved" in this way. In particle physics, many new quantities have been discovered that
obey conservation laws.

Another conserved quantity is *baryon number,* which is the
number of heavy particles participating in a reaction. The baryon
number of the proton is +1 and the baryon number of the lambda
is also +1. The other particles in the reaction have baryon
numbers of 0, because the pion and kaon are not heavy particles.
Therefore the total baryon number at the start of the reaction is
+1 and the baryon number at the end of the reaction is +1.

Physicists also defined a *strangeness* quantum number.
The proton and the pion, being nonstrange particles, each have
a strangeness of zero. On the other hand, the kaon has a strangeness of +1 and the lambda has a strangeness of −1. In this chapter, we will work with the strangeness, without stopping to explain

the mathematical details of how it is calculated. The important thing to remember is that a particle that is formed only by the strong force, yet decays by the weak force, is "strange" and has a nonzero strangeness number.

In the reaction of Fig. 24, we can see that strangeness is conserved, because the original two particles each have a strangeness of zero, while the resulting particles have their +1 and −1 strangeness numbers, adding to zero. So, in Fig. 24, a strong reaction, we see the conservation of strangeness.

What becomes of the particles formed in this reaction? The decay of the lambda particle is a weak process. About 64% of the time it decays into a proton and a negative pion, the same two kinds of particles that gave birth to it:

$$\Lambda \quad \rightarrow \quad p^+ \quad + \quad \pi^-$$

lambda   proton   negative pion

Figure 25

About 36% of the time it breaks instead into a neutron and a neutral pion:

$$\Lambda \quad \rightarrow \quad n \quad + \quad \pi^0$$

lambda   neutron   neutral pion

Figure 26

There are very rare events that occur less than one time in a thousand. Sometimes the lambda breaks into a proton, an electron, and an antineutrino. Or it breaks, even less often, into a proton, a muon, and an antineutrino.

All of these are weak reactions. The quantity known as strangeness is not conserved in these reactions. The lambda, as we have seen, has a strangeness of −1. But it decays into non-strange particles. So, in each of these decays, the total strangeness starts at −1 and changes to 0.

Physicists sum up these results by saying that strangeness is conserved in strong interactions but not in weak interactions.

"kinetic energy," the energy of motion. At the bottom of the hill, if it crashes into a fence, the collision spends the energy in breaking the fence and in the heat and sound waves that are given off. The total energy never changes.

Mass-energy

In modern physics, the mass and energy conservation laws have to be merged, because mass has been shown by Einstein to be a special "lumpy" form of energy. The mass of a subatomic particle can be converted into energy or vice versa in nuclear reactions. But a formula for the total combination of mass and energy shows that "mass-energy" is conserved.

Charge

If a glass rod is rubbed with a piece of silk, the glass rod becomes positively charged, and the silk becomes negatively charged. Their charges are equal, opposite, and add to zero . . . the same as they did when both of them were uncharged.

THE
POSITRON

17 THERE is a device about the size and shape of a kettle-drum that has played a very important part in the history of physics. This device is called a cloud chamber. In a cloud chamber, water vapor condenses along the path of a fast-moving particle. For instance, if an electron travels through the cloud chamber, its trail can be marked by a thin line of water vapor. If the electron's path is curved by a magnetic field, the vapor lies along that curve. In general, the vapor makes it possible to see the path of a particle very exactly. A scientist can see this directly by looking through a small window. More often a camera photographs the vapor trails left by particles.

The cloud chamber, invented by C. T. R. Wilson in 1911 and since improved by other physicists, was obviously a useful and fascinating piece of equipment from the very start. It gives us a chance to see the path followed by particles that were themselves too small to be seen. But nobody realized, in 1911, what surprising events were to be seen through the windows of cloud chambers.

One of the most important of these surprises was the discovery of a particle called the *positron*. The discovery of the positron took place in 1932, after twenty-one years of cloud-chamber research.

However, before we get to the discovery of the positron we have to backtrack to discuss another subject. We first have to talk about cosmic rays, for it was in studying cosmic rays that a physicist stumbled upon the track of the positron.

Cosmic rays are a kind of radiation that floods the entire universe. It is a radiation made up of very-high-energy particles of various kinds. When one of these particles comes from outer space into the earth's atmosphere, it breaks up atoms that it hits, creating an entire shower of *secondary particles*, somewhat in the way that a pebble thrown into a pond creates a shower of water droplets.

By the time the cosmic-ray showers reach the earth, they are thinned out considerably. A patch about one square centimeter in area is hit by a secondary particle only about once a minute. A *primary particle*—one that comes from outer space to start a shower—does not generally reach the earth. Long before that time, its energy has been spent.

Ever since the early years of this century, physicists have been exploring the nature of the cosmic-ray particles. One of the instruments they have used is the cloud chamber.

In 1932, Carl Anderson, the man who would later discover the muon, was studying cosmic-ray paths through a cloud chamber. The tracks that he was watching were those of secondary particles. As was common in cosmic-ray research, he was using a powerful electromagnet to make charged particles travel in curved paths.

The curvatures, intensities, and lengths of the various trails enabled him to identify the different kinds of secondary particles. He recognized the usual fragments of broken atoms, with electrons being the most common of these.

Then he noticed an interesting thing. Some of the photographs showed electron trails curving in the wrong direction. This was an unusual observation. A magnetic force field makes an electron curve in a definite, predictable direction. There was nothing known in physics that would make an electron curve backward in a magnetic field.

Anderson made another test which proved that the particles that made these backward curves could not be electrons at all. They had to be strange new particles with the same mass as an electron and the same quantity of charge as an electron, but with a charge that was opposite to that of an electron. Instead of having a negative charge, the new particle had a positive charge. In other words, it was, for all practical purposes, a positively charged electron, or, as it soon was named, a positron.

To discover such a thing was something like discovering a square drop of water. These positive electrons couldn't even exist in the electron shells of an atom. They would be forced away by the positively charged nucleus. Then, where did they come from and what part did they play in the structure of matter?

The answer to this mystery happened to be available. These positive particles had been predicted four years earlier in a strange theory advanced by a young English physicist, Paul Dirac.

Dirac's idea was that "empty space" is actually packed with electrons. The only electrons that we see are electrons that have positive energies, that is, energies greater than zero. Dirac suggested that most electrons exist invisibly at negative-energy levels. An electron at a negative-energy level would have less than zero energy. This was as strange as saying that something has less than zero mass or less than zero length, or that an event takes place in less than zero time.

However strange it seems, to understand Dirac's theory it is necessary to picture these swarms of electrons huddling invisibly in these levels of "pockets" of negative energy. Dirac imagined that if energy were provided to boost an electron out of a negative-energy pocket, it would appear as a regular electron with energy greater than zero.

Strangest of all, the empty pocket should also become detectable. It would move along, acting just like a particle. And, because it would be a moving "place" that was missing an electron, it would seem like the opposite of an electron; it would have a positive charge. In other words, it is the thing that we call a positron. As a result of Dirac's theory, positrons are often called *electron holes*, or sometimes they are just called *holes*.

A few years after the discovery of the positron, another

important experimental gain took place. The French scientist Frédéric Joliot-Curie and his wife, Irène Joliot-Curie, found that some radioactive atoms were ejecting positrons from their nuclei. Even though positrons could not exist in the electron shells, they apparently could come popping out of the nuclei. As a result of this finding, positrons were suddenly plentiful in every advanced-physics laboratory where it was possible to study their behavior in detail.

It was observed that a positron has a very short lifetime. Within a small fraction of a second, the positron meets a passing electron and the two destroy each other. The positron and the electron disappear, and a high-energy photon suddenly appears. The photon has all the energy of the two vanished particles. It is as if these twin particles with their opposite charges suddenly combine, losing all their mass and their separate identities to become a single photon. Because the two particles are completely destroyed, this event was called *annihilation.*

In other experiments, a different thing was observed. A high-energy photon could disappear, turning into an electron and a positron. This mysterious creation of matter was named *pair production.*

Here, then, were two opposite phenomena. In annihilation, two particles disappear by turning into a flash of radiation. In pair production, a flash of radiation disappears by turning into two particles.

These events matched Dirac's theory very well. Pair production happens when a photon hits an electron in a negative-energy pocket. All of the energy of the photon is transferred to the electron, so the photon vanishes, leaving an electron and a

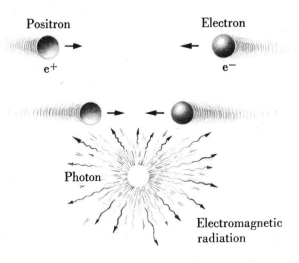

Positron

Electron

e$^+$

e$^-$

Photon

Electromagnetic
radiation

Figure 27

The annihilation process: e$^+$ + e$^-$ → γ + γ for a slowly moving positron
and electron

hole behind. Then the electron and the hole dart away just like
two ordinary particles. When the electron was in a hole, neither
the electron nor the hole could be detected; as soon as it was
knocked out of the hole, both the electron and the hole could be
detected.

How is annihilation described in Dirac's model? It is
explained like this: An electron drops into a hole. The electron
and the hole are no longer detectable. All of the energy of the
electron gets lost, and lost doubly, because the electron's energy
not only drops to zero, it drops below zero to the negative-
energy level of the hole. All of this lost energy appears as a
high-energy photon.

· · · 115

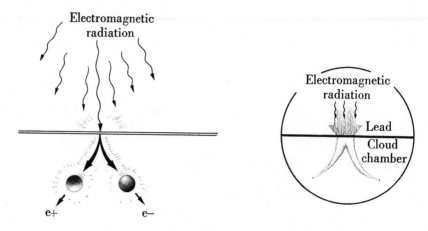

Figure 28

Electromagnetic radiation causes pair production to occur in dense materials such as lead. The resulting pair of particles can then be seen as tracks in a cloud chamber or other detector.

Dirac's imaginative model provides a very good explanation of pair production, annihilation, and the existence of positrons in general. But it is not the only workable model. There is another model that is just as good. It was devised by the American physicist Richard Feynman. According to Feynman, a positron is an electron moving backward in time.

The idea of something traveling backward in time is difficult to imagine. Watching time flow backward would be something like watching a motion-picture film being played backward. In a backward film, divers leap out of the water and onto the diving board, fires turn ashes to paper, balls roll uphill, and fruit pickers put oranges on trees.

In the world of fundamental particles, the view of time as running backward is a useful model, which describes the behavior of electrons and positrons very well. Whether time *really* runs backward or not at a subatomic level is an interesting question. But physicists are able to use this idea as a powerful and useful model, whether or not it is "true," just as an artist can paint limp watches draped like pancakes over the branches of trees whether or not watches of this kind really exist. In this way, a physicist is very much like an artist. His model can be as imaginative as he wants it to be, although it must fit the strict requirements for models described on p. 4.

In Feynman's model an electron–positron pair is what we see when a single electron is moving back and forth through time. Our view of this phenomenon is limited, since our own time flows in only one direction, from past to future. Therefore it would look to us like two opposite particles (a positron and an electron) existing at the same time.

To picture what happens, imagine that these events are happening very slowly, over a period of minutes, instead of happening as they actually do in an infinitesimal fraction of a second. Suppose, then, that an electron is moving along in a usual sort of way, going forward in time starting at 12:00 noon. Then suppose that at 12:05 the electron changes its direction in time. Instead of continuing through later and later times (12:06, 12:07, etc.), it reverses itself and goes back through time to 12:04, 12:03, and so forth, all the way back to 12:00 noon. The backward travel through time would make the electrical charge on the electron appear to be positive instead of negative. It would look like a positron.

How would all this appear to a human observer? At 12:00 noon he would seem to see two particles. But it would really

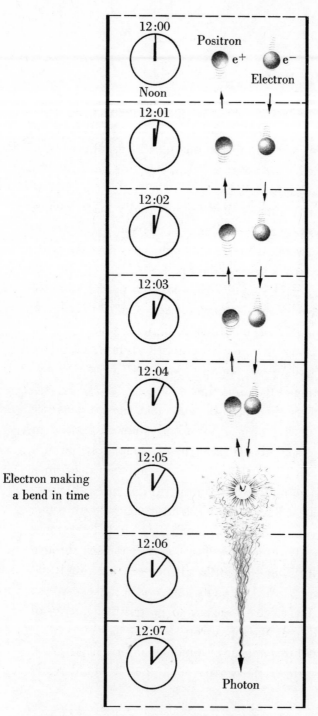

Figure 29

Feynman's time-travel model; annihilation occurs at 12:05 when the electron makes its time reversal.

be the same electron moving forward in time and moving backward in time. We know that the electron was there at 12:00 noon to begin with, going forward in time. We also know that it was there at 12:00 noon going backward in time, having "turned around" at 12:05.

If the observer continues to watch, he sees the "two" particles coming closer and closer together, until, at 12:05, they meet and disappear.

After 12:05 no particle is seen. We know that the electron, having "turned around in time" at 12:05, did not go on into 12:06, 12:07, etc. What seems to the observer like two particles annihilating each other is only a single particle making a bend in time.

Also, at 12:05, the electron had to give off energy, a sort of recoil, in order to make its time-swerve. That energy appears as a photon. The observer, continuing to go forward in time (as people have to do), sees the photon as the only thing remaining at 12:06, 12:07, and later times.

Pair production can be seen in a similar way. There we *start* with an electron going backward in time, looking like a positron. It is hit by a photon, and the energy of the impact makes it reverse and go forward in time. An observer would see a photon disappear and "two" particles appear at the moment of time reversal. Once more, of course, he is merely seeing a single particle making a bend in time.

Even though time-traveling electrons are hard to imagine, Feynman's time-travel model is as exact and as useful as Dirac's negative-energy levels. The two models predict exactly the same kinds of behavior among the particles.

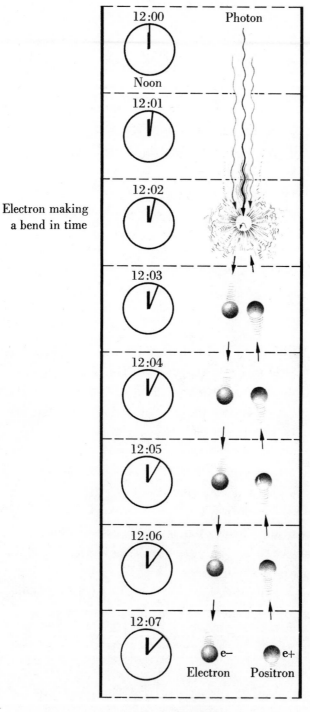

Figure 30

Feynman's time-travel model; pair production occurs at 12:02 when the electron makes its time reversal.

The fact that two different models can so neatly and beautifully describe the same events indicates something in general about the models used by physicists. Models, no matter how much precision and imagination goes into their making, are merely pictures and descriptions of a strange, ultimate reality that we can never totally know.

ANTIMATTER

18 POSITRONS are holes left by electrons departing from negative-energy levels. We can also see them as electrons traveling backward in time. In either case, the positron is the opposite of an electron, a sort of "reverse electron," or "antielectron."

Having discovered positrons, physicists began to speculate about other kinds of reverse particles. In fact, it was possible to imagine a world made of *antimatter*, where positrons rippled in Bohr orbits around nuclei containing negatively charged antiprotons.

Antiparticles do exist. After physicists found the positron, they set out to find the *antiproton*. Gamma rays of very great energy are needed in order to produce antiprotons. Depending on how you look at it, the production of antiprotons means either boosting a proton out of a negative-energy level or jolting a proton to make it go backward in time. Since a proton is about 1,840 times as massive as an electron, the gamma rays that are needed for this task have to be 1,840 times as energetic as the gamma rays that produce positrons.

So physicists had to wait until extremely high-energy particle accelerators could be built before they could generate antiprotons. When these accelerators were built, the expected results did happen. The paired tracks of massive particles curving in reverse directions were seen in detectors. The antiproton had been found.

When a proton and an antiproton meet, they annihilate each other, just as an electron and a positron do. However, in the case of the proton and antiproton, the gamma rays produced are much more energetic.

Various experimental and theoretical proofs eventually showed that every particle has its antiparticle. But, you might ask, what about the neutron? Since a neutron has no charge, how can there be an *antineutron?*

It turns out that there is another way in which a particle and an antiparticle can be opposite, other than in what their electric charges are. They can also be opposite in their magnetic fields. The magnetic field of the neutron is oriented in the opposite way. So even the uncharged neutron and antineutron can be opposites. When a neutron and an antineutron meet, annihilation occurs and the two particles disappear, giving birth to photons, which carry away the energy.

In some cases, an uncharged particle is its own antiparticle. This is true for photons and also for the neutral pion.

Figure 31 is a chart of all the particles that were known in 1947. The particles are on the left side of the chart, and their antiparticles are shown to the right.

PARTICLES	ANTIPARTICLES
BARYONS $p+$ n PROTON NEUTRON (nucleons)	$\overline{n}$ $\overline{p}-$ ANTINEUTRON ANTIPROTON
MESONS $\pi+$ π° POSITIVE PION NEUTRAL PION	$\pi-$ NEGATIVE PION
LEPTONS $\mu+$ POSITIVE MUON $e-$ ELECTRON ν_μ MUON NEUTRINO ν_e ELECTRON NEUTRINO	$\mu-$ NEGATIVE MUON $e+$ POSITRON $\overline{\nu_\mu}$ MUON ANTINEUTRINO $\overline{\nu_e}$ ELECTRON ANTINEUTRINO

γ
PHOTON

Figure 31

PARTICLES CHART

The chart also divided the particles (and their antiparticles) into different sets based on their masses. The largest, most massive were called *baryons,* the smallest, least massive were called *leptons,* and those that were in between were called *mesons.* The heavy baryons included protons, neutrons, and their antiparticles. The mesons included only the various kinds of pions. Muons, electrons, positrons, and neutrinos fitted into the class of leptons. Photons were an exception, and fitted into none of these three classes.

Notice that the muons, even though they were originally thought to be mesons, have been put in with the leptons. That is because they act like electrons, and obey a law having to do with "conservation of lepton number." In fact, that law would break down in reactions involving both electrons and muons, if muons were not considered as leptons. So, we see a great variation in mass among the leptons, since muons are 207 times as massive as electrons, while neutrinos have no mass at all.

Since pions are all by themselves as mesons, why have a class of mesons? The answer is that the pions did not stay alone very long. Other mesons were to be discovered, as is shown in a later chapter. Gradually physicists were to learn what it meant for a particle to be a meson.

Figure 31 is useful in getting a feeling for the world of particles. It presents a fairly simple picture of a world made up of only sixteen kinds of particles. However, it was an oversimplified picture. Many other particles were soon to make their appearance.

Meanwhile, even in the sixteen-particle days, there were

many problems facing physicists. One of these was the puzzle of the muon, and its place in the structure of the universe. Another was the puzzle of antimatter.

The equations and theories of physics implied that there should be roughly equal quantities of particles and antiparticles in existence. Nevertheless, we live in a world made of particles, where antiparticles are merely rare visitors conjured up in special experiments to live very brief lives before being annihilated.

It would be impossible to live in a world made up of both matter and antimatter. If a lump of antimatter with the mass of an apple were to come into contact with matter, the lump of antimatter and an equally large lump of matter would annihilate each other, dissolving into a burst of electromagnetic energy. That would happen because all the antiprotons and antineutrons and positrons in the antimatter apple would meet the protons, neutrons, and electrons of the surrounding matter, and all of these particles would annihilate each other. The resulting burst of energy would be far greater than any man-made nuclear explosion.

It's obvious why we don't live in a world of mixed matter and antimatter. Such a world would blow itself apart. But the question still arises: Where is the antimatter half of the universe? Where do the antiparticles come from?

One theory starts with the origin of the universe. At the beginning of time, when our universe began as a vast explosion, there was a flood of electromagnetic energy rushing outward. This deluge must have generated pairs of particles and antiparticles in huge numbers, and that was how matter was formed. The particles eventually formed the planets and stars and other matter that we are familiar with. But where did the anti-particles go?

Perhaps some entire sectors of our universe are made of antimatter. Perhaps there are star clusters and galaxies made of antimatter. It is not obvious whether a distant galaxy is made of matter or antimatter, because our knowledge about distant parts of the universe comes to us from photons—X rays, ultra-violet rays, light, infrared waves, and radio waves. Since a photon is its own antiparticle, the same photons would come from stars of antimatter that come from stars of matter.

There is one way that distant stars act differently, depend-ing on whether they are made of matter or antimatter. Our sun emits floods of neutrinos, and so would every star made of matter. An antimatter star would emit floods of antineutrinos.

If our universe is made totally of matter, then a continuous "rain" of neutrinos must be flowing through space. If our universe is a balanced mixture of matter and antimatter, with some sections mainly matter and some sections mainly anti-matter, then there must be a balanced flow of both neutrinos and antineutrinos through space.

The universal background of neutrinos (or of neutrinos and antineutrinos) is very faint and hard to detect. Therefore, we are still not able to measure this universal neutrino back-ground to find out whether we live in a mixed matter and anti-matter universe or simply a matter universe. However, if neu-trinos bring us the answer that we live in a universe of matter, it would seem that the antiparticles formed at the beginning of time must have formed a whole different universe—a different space and time, an antiuniverse existing like some strange fan-tasy on an entirely different level of reality.

A MENAGERIE OF PARTICLES

19 MANY other nonstrange particles have been discovered in recent years. Arising in strong interactions and decaying in strong interactions, these particles exist for such extremely short lifetimes that they are not included on our list of stable and semistable particles. Their lives are about one hundred-thousandth of a billionth of a billionth of a second in duration.

Figure 32 shows the stable and semistable particles known in 1973, including the various strange particles discovered during the previous twenty years. This table is an expanded version of the table shown on p. 124, when the number of particles was far smaller.

Some general terms appear in Fig. 32, which are helpful in establishing some order among this bewildering array of particles. Here are some useful definitions, most from earlier in this book, but some are presented here for the first time:

Baryon—a heavy particle.

Nucleon—a proton or neutron.

Hyperon—a baryon that is heavier than a nucleon.

Meson—a particle of medium mass.

Hadron—any particle that can take part in a strong interaction. This includes all baryons and mesons.

Lepton—a particle of low or zero mass that can experience electromagnetic forces or the weak force, but is never involved in strong interactions.

Boson—a particle with a spin of 0 or 1 that does not obey Pauli's exclusion principle. That is, there is no upper limit on the number of bosons with identical quantum numbers that can be packed into the same space. This includes photons and all mesons. It is interesting to notice that all particles that act as the "carriers" of a force (pions, photons, and the W-particles—if they exist) are bosons.

Fermion—a particle with a spin of $\frac{1}{2}$ or $1\frac{1}{2}$ that does obey Pauli's exclusion principle. This includes all baryons and leptons.

· · · 129

OMEGA MINUS $\Omega-$	**ANTIOMEGA PLUS** $\overline{\Omega+}$	
charge = −1	charge = +1	
baryon # = +1 mass =	baryon # = −1 mass =	
strangeness = −3 1672 MeV	strangeness = +3 1672 MeV	
XI MINUS $\Xi-$	**ANTIXI PLUS** $\overline{\Xi+}$	
charge = −1	charge = +1	
baryon # = +1 mass =	baryon # = −1 mass =	
strangeness = −2 1321 MeV	strangeness = +2 1321 MeV	
XI ZERO $\Xi°$	**ANTIXI ZERO** $\overline{\Xi°}$	
charge = 0	charge = 0	
baryon # = +1 mass =	baryon # = −1 mass =	
strangeness = −2 1315 MeV	strangeness = +2 1315 MeV	
SIGMA MINUS $\Sigma-$	**ANTISIGMA PLUS** $\overline{\Sigma+}$	
charge = −1	charge = +1	
baryon # = +1 mass =	baryon # = −1 mass =	
strangeness = −1 1197 MeV	strangeness = +1 1197 MeV	
SIGMA ZERO $\Sigma°$	**ANTISIGMA ZERO** $\overline{\Sigma°}$	
charge = 0	charge = 0	
baryon # = +1 mass =	baryon # = −1 mass =	
strangeness = −1 1192 MeV	strangeness = +1 1192 MeV	
SIGMA PLUS $\Sigma+$	**ANTISIGMA MINUS** $\overline{\Sigma-}$	
charge = +1	charge = −1	
baryon # = +1 ` mass =	baryon # = −1 mass =	
strangeness = −1 1189 MeV	strangeness = +1 1189 MeV	
LAMBDA Λ	**ANTILAMBDA** $\overline{\Lambda}$	
charge = 0	charge = 0	
baryon # = +1 mass =	baryon # = −1 mass =	
strangeness = −1 1115 MeV	strangeness = +1 1115 MeV	
NEUTRON n	**ANTINEUTRON** $\overline{n}$	
charge = 0	charge = 0	
baryon # = +1 mass =	baryon # = −1 mass =	
strangeness = 0 939½ MeV	strangeness = 0 939½ MeV	
PROTON p+	**ANTIPROTON** $\overline{p-}$	
charge = +1	charge = −1	
baryon # = +1 mass =	baryon # = −1 mass =	
strangeness = 0 938¼ MeV	strangeness = 0 938¼ MeV	

Left side vertical labels: HADRONS, BARYONS, HYPERONS, NUCLEONS

Right side vertical label: FERMIONS

Figure 32

PARTICLES CHART

	PARTICLES		ANTIPARTICLES	

<table>
<tr><th colspan="2">PARTICLES</th><th colspan="2">ANTIPARTICLES</th></tr>
</table>

PARTICLES | **ANTIPARTICLES**

	PARTICLES	ANTIPARTICLES
HADRONS / MESONS (BOSONS)	ETA η charge = 0 strangeness = −1 mass = 549 MeV	
	KAON ZERO K° charge = 0 strangeness = +1 mass = 498 MeV	ANTIKAON ZERO $\overline{K^\circ}$ charge = 0 strangeness = −1 mass = 498 MeV
	KAON PLUS K^+ charge = +1 strangeness = +1 mass = 494 MeV	KAON MINUS $\overline{K-}$ charge = −1 strangeness = −1 mass = 494 MeV
	PION PLUS $\pi+$ charge = +1 strangeness = 0 mass = 140 MeV	PION MINUS $\pi-$ charge = −1 strangeness = 0 mass = 140 MeV
	PION ZERO π° charge = 0 strangeness = 0 mass = 135 MeV	
LEPTONS (FERMIONS)	MUON MINUS $\mu-$ charge = −1 lepton # = +1 mass = 106 MeV	MUON PLUS $\mu+$ charge = +1 lepton # = −1 mass = 106 MeV
	ELECTRON $e-$ charge = −1 lepton # = +1 mass = ½ MeV	POSITRON $e+$ charge = +1 lepton # = −1 mass = ½ MeV
	MUON NEUTRINO ν_μ charge = 0 lepton # = +1 mass = 0	MUON ANTINEUTRINO $\overline{\nu_\mu}$ charge = 0 lepton # = −1 mass = 0
	ELECTRON NEUTRINO ν_e charge = 0 lepton # = +1 mass = 0	ELECTRON ANTINEUTRINO $\overline{\nu_e}$ charge = 0 lepton # = −1 mass = 0
BOSON	PHOTON γ charge = 0 mass = 0	

NOTE: the masses of the particles in this table are expressed in energy units called "MeV," standing for "millions of electron volts." One MeV is the energy that an electron would lose or gain in going through a one-million-volt electrical force field.

QUARKS

20 PROBING at nucleons with high-energy beams, physicists in the late 1960s were able to detect a graininess to the proton and the neutron. Each of these particles seemed to have three separate centers of charge inside it. In the case of the neutron, the charge centers would have to balance out to add to a total charge of zero. In the proton, they would combine to form a single quantum of charge. The interesting possibility arising from these findings is that nucleons might not be basic particles, but instead might be formed from particles that are even more fundamental. These underlying centers of charge were temporarily named "partons."

Meanwhile, an attack on the possible substructure of particles in general was being made from a different direction. A leading physicist at the California Institute of Technology, Murray Gell-Mann, was studying the entire list of known particles and was beginning to see a pattern. He saw that a logically simple scheme could be constructed using three mysterious units. He named these units *quarks*. And to his collection of three quarks, he added three *antiquarks*, to make a total of six mysterious units.

At first, Gell-Mann did not think of his quarks and antiquarks as particles and antiparticles. Instead, he thought of them as mathematical rules of existence that lay below the particle world. But, eventually, he and other physicists did begin to think of them as particles. In fact, they were probably the grainy "partons" that had been detected by experimental beams. But the name quark had won out, and when physicists talk about these subparticles, they no longer call them partons. They call them quarks.

Gell-Mann divided his three quarks into *flavors*. He used the word "flavor" more or less playfully, and not having any connection at all with the usual meaning of the word. And in the same casual style, he called the three quark flavors "up," "down," and "sideways," abbreviated as u, d, and s. Later, it began to be obvious that the sideways flavor had to do with strangeness, so the three flavors became "up," "down," and "strange." But the letters u, d, and s stayed the same.

Gell-Mann speculated that the three quarks would have the properties listed in Fig. 33:

Quark flavor	Baryon number	Charge	Strangeness	Spin
u	⅓	+⅔	0	½
d	⅓	−⅓	0	½
s	⅓	−⅓	−1	½

Figure 33

The three quarks

He pictured every baryon to be made up of three quarks. A proton, for example, would consist of quarks *u, u, d*. The charges on these three quarks would add to +1, the baryon number would add to +1, and the strangeness would add to 0. Two of the half-spins line up in opposite directions to cancel each other out, leaving the half-spin of the third quark as the spin of the proton. This arrangement fits the proton's actual properties.

A neutron, on the other hand, would consist of quarks *u, d, d*. This time, the charges would add to zero, but all other characteristics would be the same as for the proton.

Figure 34 shows the makeup of various baryons, each one being made up of three quarks. Going back to Fig. 33, notice how each baryon gets its characteristics from its three quarks. For instance, the omega (Ω) is a baryon with a strangeness of −3, and a charge of −1. You can see how the three *s* quarks lead to the right results for the omega.

Omega (Ω) Antiomega $(\overline{\Omega})$

s s s   *s s s*

Xi minus $(\Xi-)$ Antixi plus $(\overline{\Xi+})$

d s s   *d s s*

Xi zero (Ξ°) Antixi zero $(\overline{\Xi^\circ})$

u s s   *u s s*

Sigma minus $(\Sigma-)$ Antisigma plus $(\overline{\Sigma+})$

d d s   *d d s*

Sigma zero (Σ°) Antisigma zero $(\overline{\Sigma^\circ})$

u d s   *u d s*

Sigma plus $(\Sigma+)$ Antisigma minus $(\overline{\Sigma-})$

u u s   *u u s*

Lambda (Λ) Antilambda $(\overline{\Lambda})$

u d s   *u d s*

Neutron (n) Antineutron $(\overline{\mathsf{n}})$

u d d   *u d d*

Proton $(\mathsf{p}+)$ Antiproton $(\overline{\mathsf{p}-})$

u u d   *u u d*

Figure 34

Each meson, on the other hand, is made of a quark and an antiquark. For instance, a positive kaon (K^+) has quark structure *s u*, to give it its strangeness of +1, its baryon number of 0, its spin of 0, and its +1 quantum of charge.

In general, then, there are three ways to combine quarks. Three quarks make a baryon. Three antiquarks make an antibaryon. A quark and an antiquark make a meson. From these different arrangements all known *hadrons* (all particles that feel the strong force) can be constructed.

$\cdots$ 135

There was a small problem in this neat scheme. Each quark has a half-spin. A particle with a half-spin is a *fermion* and must obey the Pauli exclusion principle. That is, two identical particles with identical sets of quantum numbers cannot be in the same atomic or subatomic system together. Yet, many particles obviously contain quarks of identical flavors. Three *s* quarks make up an omega particle. Two *u* quarks and a *d* quark make up a proton. And we have seen several other examples of quark duplication. All baryons would have this trouble except one with a *u d s* arrangement or an antibaryon with *u d s*. And physicists definitely were not interested in questioning whether or not every particle with half-spin was a fermion and therefore subject to the exclusion principle. Treating half-spins that way was one of the most important ideas in physics, and the whole structure of particle physics would be badly shattered if that idea were discarded.

Instead the American physicist Sheldon Glashow suggested that quarks had another quality besides flavor. He called this quality *color*. As with flavor, the color of a quark has no connection with the usual use of the word *color*. And when Glashow declared that every quark or antiquark was either red, yellow, or blue, he was using those words as arbitrary labels.

But now the three *s* quarks that make up an omega particle are all different from one another. One is red, one yellow, and one blue. And, in general, every baryon (and antibaryon) fits this rule. Their three quarks (whether they come in identical flavors or not) have to be of three different colors.

This mix of three colors results in baryons being "colorless." That is, the red, yellow, and blue balance, so that there is no leftover redness, blueness, or yellowness in any baryon. It is important that this new quantum condition balance out in this way, to fit the fact that the known baryons do not come in additional forms. If all different color combinations were allowed, there would have to be ten different kinds of neutrons, ten different kinds of sigma-plus particles, and, in general, ten varieties of every existing baryon and antibaryon. The color-balancing rule, with its absolute insistence on red, yellow, and blue, agrees with the fact that those sets of ten are not observed.

Nobody has any idea what "color" would be like if it were observed, since nobody knows what it actually is. But whatever it is, it is not making an appearance. So it makes sense to believe the red-yellow-blue rule for baryons.

But what about mesons? A meson has only two quarks. For example, if one is red and one is blue, then "color" would appear in that meson as "nonyellowness," and in some way nonyellowness, whatever it is, would manifest itself in particle physics.

The answer that was suggested is: First, let the quark and antiquark in a meson be of the same color. This causes no trouble for the exclusion principle, because the quark and antiquark automatically have different quantum states and therefore don't need to be of different colors. Instead, they change colors together at a very swift rate. In one instant they are red. At another instant they are blue. At still another instant they are yellow. They change color so fast that their swift-changing colors balance out, and mesons are colorless after all.

So, there are two ways that hadrons stay colorless. One way, among the baryons, is to mix the three quark colors at one

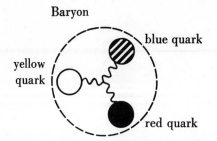

Baryon

yellow
quark

blue quark

red quark

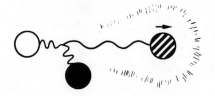

blue quark knocked out of baryon
by high-energy photon

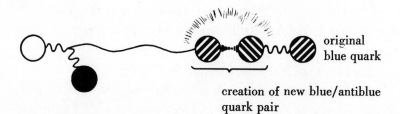

original
blue quark

creation of new blue/antiblue
quark pair

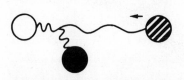

newly-formed blue quark is pulled
back by red and yellow quarks to
rebuilt baryon

newly-formed antiquark joins out-
ward-bound quark to form a
meson

Baryon

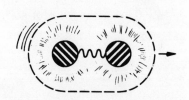

the quark and antiquark in this
meson change colors rapidly,
flickering between blue, red, and
yellow

Figure 35

time. The other way, among the mesons, is to change quark colors very rapidly.

As to leptons and photons, there is no indication that they have any internal structures. It seems as if electrons, positrons, muons, neutrinos, and photons are truly basic particles, not made of quarks or anything else. Electrons have been probed with very-short-wavelength X-ray beams and no trace of an internal structure has been detected. Only the hadrons fit the quark theory.

What does quark color do, besides keeping quarks in different quantum states? It plays no part at all in how particles appear to us. All the properties of mass, charge, spin, strangeness, and so forth that give hadrons their wide range of observable qualities are based on the *u d s* flavors, not the red, yellow, blue colors.

Glashow thinks that color may have another very important part to play. He has suggested that quark color has to do with the strong force. The quarks exchange particles called *gluons,* and as they carry out this exchange, the quarks trade colors. So that even the three quarks in a baryon change colors. Perhaps the one that was blue a moment ago is yellow now, and the one that was yellow is now blue. In Glashow's theory, the exchange of gluons (and of color) among the quarks is the real source of the strong force. The strong force that operates between hadrons, as virtual pions go back and forth, is only a faint echo of the force between the quarks.

THE BOTTOM OF THE BARREL

21 HAVING arrived at the idea of quarks, do physicists expect some still more basic level of reality? We have come down through several levels already. First there were atoms. Then the atom was broken into electrons and a nucleus. Then the nucleus was broken into protons and neutrons. Now protons and neutrons are thought to be made of quarks. We are at the fourth level already. Is there a fifth level? A sixth? An infinite number of levels?

Some physicists think that quarks are the end of the story and that things break down no further. To back up this idea they use something called the *quark confinement hypothesis.* This hypothesis states that it is impossible to isolate a quark.

What they think happens with quarks is this. Suppose that a high-energy photon knocks a blue quark out of a baryon. The energy that it takes to do this is so enormous that, besides the blue quark that is broken loose, a blue–antiblue quark pair is produced, in a very-high-energy version of the way that electron–positron pairs are produced. The newly created blue rushes into the empty spot in the broken nucleus. This keeps the orignal baryon in existence. The newly created antiblue joins the loose blue to form a quark–antiquark arrangement—a meson. Then, all that the observer sees is that the baryon has ejected a meson. Because of this process, no loose quark could ever exist.

This quark confinement hypothesis goes on to state that there is no way to investigate or conjecture about the internal structure of a particle that you can't even isolate to begin with. If no quark can ever be produced then it is meaningless to ask whether it is made of still more basic particles.

Before closing the discussion of quarks, let's come back to the quark flavors. These flavors, in their various mixtures, are what cause all the variety among hadrons. Apparently the weak force operates to change the flavors of quarks. A quark emits a W-particle, which breaks down into an electron and an antineutrino, and the quark changes its flavor.

The weak force is a link between hadrons and leptons, because it involves both of them. Various leptons collide with hadrons or are ejected by hadrons as the weak force operates.

The quark concept made the list of basic particles much smaller. Now it looked like this:

Quarks	Anti-quarks	Leptons	Anti-leptons		Force carriers
u	$\overline{u}$	e^-	e^+		gluons
d	$\overline{d}$	ν_e	$\overline{\nu}_e$		W-particles
s	$\overline{s}$	μ^-	μ^+		photons
		ν_μ	$\overline{\nu}_\mu$		

Figure 36

Particle table with three quark flavors

Neglecting the three carriers for a moment, look at the main part of the table. The particles above the heavy line are all the particles needed to construct the world. The whole material universe and its energetic processes need only the u and d quarks, the electron, the electron neutrino, and their antiparticles. The particles below the line appear only in high-energy reactions, mainly in experiments devised by physicists, but also occasionally in cosmic-ray showers.

The s quark is much heavier than the more ordinary u and d quarks. That greater mass of the s fits the high masses of the *strange hyperons* (baryons that are heavier than nucleons) discovered in recent years.

However, the two empty spaces to the lower left of the table were troublesome to Sheldon Glashow. They seemed like a break in a balanced pattern. So he predicted the existence of a fourth quark. He used the word *charm*, abbreviated by the letter *c*, as the name of this quark and its flavor. So now the table of quarks looks like this:

Quark flavor	Baryon number	Charge	Strangeness	Charm	Spin
u	⅓	+⅔	0	0	½
d	⅓	−⅓	0	0	½
s	⅓	−⅓	−1	0	½
c	⅓	+⅔	0	+1	½

Figure 37

The four quarks

And now the particle table looks like this:

Quarks	Anti-quarks	Leptons	Anti-leptons
u	$\bar{u}$	e^-	e^+
d	$\bar{d}$	ν_e	$\bar{\nu}_e$
s	$\bar{s}$	μ^-	μ^+
c	$\bar{c}$	ν_μ	$\bar{\nu}_\mu$

Force carriers
Gluons
W-particles
Photons

Figure 38

Particle table with four quark flavors

Physicists felt that, as increasingly higher-energy techniques became available, they might discover still heavier particles, giving some experimental backing to the existence of the *c* quark and its antiquark. Starting in late 1974, a series of hyperons, heavier than any of the known strange hyperons, were discovered. The first of these was named the *J-particle*, but soon a whole series of several other similar heavy particles were generated in very-high-energy collisions. They were given the name *psions*. The psions are evidence that there is a charmed quark, and that it is heavier than the *s* quark.

At the time of this writing, physicists are trying to determine whether the psions have some unusual property to represent the existence of charm. But that property has not yet been identified.

There are other questions that remain unanswered. Why do the particles below the heavy line in the table of particles exist at all? Like the muon on p. 96, the universe would do quite well without them.

Some physicists have questioned the entire quark idea. Is it really true that quarks are locked invisibly into matter in such a way that one can never be detected? Or will someone detect a quark? Physicists haven't surrendered and accepted the confinement theory, and they still are trying to detect quarks. But, so far, the quark has not been found.

Perhaps the whole system of quarks is too complicated. Perhaps the ultimate particles are fewer in number and fit together in some simpler arrangement. Or perhaps Glashow was right in his conjecture. Perhaps we are at the end of the downward spiral, and perhaps quarks are at the bottom of the barrel of particles.

CONCLUSION

22 WE have been on a long journey through a submicroscopic world. We have seen one fantastic model of reality after another. We have seen the vague, early models where atoms were nothing but infinitesimal spheres of matter. From there we went to the raisins-in-a-bun model, where the negatively charged electrons were scattered through the positively charged stuff of atoms. After that, Rutherford's nuclear atom was replaced by Bohr's "impossible atom" with its strange rules governing electron orbits. Then Schrödinger's wave-mechanical atom took over. Finally, physicists plunged down into the center of the atomic nucleus to discover still stranger and more fundamental particles. The latest models show the universe to rest on a foundation of quarks, particles so hidden that, by definition, they can never be detected.

But the changing world of physics has seen the overthrow of many theories that once were thought to represent the ultimate nature of the universe. There is no certainty that the quark model has brought us to the end of the trail. Other, still stranger realities may await us.

The world of particles is very strange, compared with the everyday world of our experience. It is strange to realize that these two worlds fit together—that our everyday reality is somehow built out of particles such as protons, pions, and quarks, things that are both particles and waves, things that might move backward and forward in time.

This peculiar behavior has led some physicists to speculate that particles may not exist in space or time, that their "real" existence is outside of space and time and that their peculiar traits are a result of the fact that space and time are, to some extent, part of our way of seeing things. In our perceptions, we organize the things that we experience, arranging events so that we think of 12:00 as coming before 12:01. We think of one object as being next to another object. We can hardly picture a universe where things aren't organized in that way. Only in our thoughts and our dreams do we come close to being free of the limitations of space and time. In a dream we can move from place to place or time to time without the usual limits or the usual logical restrictions.

Perhaps it is our own limitation in having to arrange our perceptions in space and time that clouds the universe from our view. Certainly the physical universe is much more dreamlike and much less mechanical than we generally realize. Modern physics is a whole new way to perceive the universe we inhabit. Perhaps we can eventually learn to view this universe beyond the limits of space and time.

INDEX